Fuzzy Logic

Fuzzy Logic

Edited by

J. F. Baldwin
University of Bristol

JOHN WILEY AND SONS

Chichester · New York · Brisbane · Toronto · Singapore

Other Wiley Editorial Offices

John Wiley & Sons, Inc., 605 Third Avenue,
New York, NY 10158-0012, USA

Jacaranda Wiley Ltd, 33 Park Road, Milton,
Queensland 4064, Australia

John Wiley & Sons (Canada) Ltd, 22 Worcester Road,
Rexdale, Ontario M9W 1L1, Canada

John Wiley & Sons (Asia) Pte Ltd, 2 Clementi Loop #02-01,
Jin Xing Distripark, Singapore 0512

British Library Cataloguing in Publication Data

A catalogue record for this book is available from the British Library

ISBN   0 471 96281 3

Produced from camera-ready copy supplied by the authors
Printed and bound in Great Britain by Bookcraft (Bath) Ltd
This book is printed on acid-free paper responsibly manufactured from sustainable forestation,
for which at least two trees are planted for each one used for paper production.

Contents

Preface

In 1965 Lotfi Zadeh published a paper called 'Fuzzy Sets' and later stated the principle of incompatibility:

"As the complexity of a system increases, our ability to make precise and significant statements about its behaviour diminishes until a threshold is reached beyond which precision and significance become almost mutually exclusive characteristics."

Other scientists and mathematicians, stimulated by Zadeh's vision for new ways of looking at complex systems, joined forces with him and a new body of researchers was born. In recent years, Fuzzy Logic and Fuzzy Control have been used in many real world applications. It is now an established subject in the areas of control and decision and knowledge engineering.

Many people confuse fuzzy uncertainty with probabilistic uncertainty. Do we really require another calculus for handling uncertainty which is additional to that provided by probability theory? This is an important question for any potential reader of this book. We can answer this question by considering the following two answers to the question concerning the value of a fair dice: (1) the value is even, (2) the value is large. With answer (1) we can use probability theory to say that the value is 2 or 4 or 6 with equal probability. With answer (2) we need to know what we mean by the fuzzy term 'large' before we attempt to derive a probability distribution. We could assume 'large' in this context means 5 or 6, in which case the value of the dice would be 5 or 6 with equal probability. This would be an approximation since others may prefer to say that 4 could also satisfy the concept 'large'. In fuzzy set theory we allow each integer value in the range 1 to 6 to have a membership value in the fuzzy set 'large'. The values 1, 2 and 3 could all have membership values of 0. The values 5 and 6 could have the membership value 1. The value 4 could have some membership value in the region of 0.5. This choice of membership value is once again fuzzy but we choose to give precise values in the interval $[0, 1]$. This is less approximate than choosing values of 0 or 1 for the membership value which is what we are doing when we do not accept fuzzy set theory. From these membership values for the fuzzy set 'large' we can construct a probability distribution for the value of the dice. This requires additional assumptions which we will not describe here. The important point is that we need both fuzzy and probabilistic theories if we are to use fuzzy concepts.

Why should we need to use fuzzy concepts? Why did we use 'large' to describe the value of the dice and not simply use '5 or 6'? Why does natural language use so many concepts which have no precise definition, fuzzy concepts which are allowed a certain flexibility in their interpretation? The world is continuous in a macroscopic sense. We interpret the world in terms of classifications expressed in linguistic form and invented by man. We discretize the continuous but with flexibility. We use fuzzy concepts to avoid sharp boundaries between similar cases. Some trees are definitely trees and not bushes. Others are less definitely trees and more possibly bushes.

Various shades of grey are allowed in addition to black and white. Truth can vary in intensity. There can be various degrees of guilt, of love, of cleanliness, of belief, of reliability. These degrees can be clustered into linguistic terms such as almost certainly guilty, very much in love, strong belief, very reliable. More important clustering is with multiple attribute objects such as trees, chairs, bright day, pretty girl, happy times, etc. These groupings are clusters introduced to simplify our understanding of the real world and to encourage meaningful communication between people. All the clusters represent fuzzy concepts. Remove the fuzziness and you approximate. This approximation can be very dangerous. Consider a design specification which requires a reliability of at least 0.95. Of our possible designs, one could have a reliability of 0.949 but be much better than any other design in all other respects. It would not be chosen unless we allow some flexibility in the specification. We could use 'large' instead of 0.95, and this would provide the required flexibility.

Fuzzy Set Theory provides a flexibility for representing the real world. Fuzzy Logic allows logic style statements to be made about the real world using this flexible form of representation, and provides methods for deriving inferences from these statements. Fuzzy Logic is a logic allowing shades of grey, degrees of in betweens which provides flexibility in knowledge representation and a robustness in the methods of inference.

An important part of the inferential robustness is the ability of fuzzy logic to generalize from prototypical cases to new situations. This process of induction is extremely important in human intelligence. We see a new object in a room and conclude that it is a chair even though we have never seen such an object before. It seems to have the required strength to hold a person even though the seat is a most peculiar shape, it has a back which appears that one could lean against even though it is different in design from any other chair seen in the past, it is in a position that one would expect a chair, *etc.* We induce that this is a chair. Fuzzy Sets can be used in constructing a theory of generalization, a theory of case based reasoning. This theory will be a method of interpolating from similar but known situations. There is no rational justification of induction like there is for deduction. We cannot claim that a given inductive inference is valid. No method of induction can be validated. Nevertheless, we find in practice that many applications requiring some form of induction seem to benefit from methods using fuzzy sets.

We have attempted to justify or at least indicate why fuzzy sets and fuzzy logic are important for many real world application areas. The subject area is young and there is still much to be done with the establishment of the fundamentals and development of methods for applications. In particular, one would hope that the future will find a greater exploitation of fuzzy sets in linguistics. Natural language interpretation and representation is important in computer science to allow easy communication between computers and humans. Fuzzy Sets could well play an important part in future developments in this area.

This book results from collecting together extended papers form those presented at the Neuro Fuzzy Conference held at Brunel in 1995. The conference dealt with Soft Computing, a term coined by Zadeh, to bring together methods in the areas of probability theory, fuzzy sets, neural computing and genetic programming. It is early days to see exactly the relationship between these approaches, but mixtures of these methods are being used in applications.

The papers in this volume illustrate and extend many of the points mentioned in this preface. There are papers which deal with the fundamentals of possibility theory, with the processes of generalization and induction, with inference. There are papers which deal with induction and data mining. The relationship of fuzzy sets with natural language is also discussed. Other papers deal directly with real applications and form an important illustration of the use of fuzzy sets in engineering and other application areas. There are papers which use a mixture of fuzzy set theory and neural net theory. The reader will be introduced to many new ideas in fuzzy set theory and will find interesting applications.

J. F. Baldwin

University of Bristol

List of Figures

List of Tables

List of Contributors

Baldwin J.F., University of Bristol, Department of Engineering Mathematics, Queen's Building, University Walk, Bristol BS8 1TR, UK.

Bennett S.C., School of Computing Sciences, De Montfort University, The Gateway, Leicester LE1 9BH, UK.

Bersini H., IRIDIA, Université Libre de Bruxelles - CP 194/6, 50, ave. Franklin Roosevelt, 1050 Bruxelles, Belgium.

Carvalho M.T., CVRMUTL - Departamento de Engenharia de Minas, Instituto Superior Técnico, CVRM - Centro de Valorização de Recursos Minerais, Av. Rovisco Pais, 1096 Lisboa Codex, Portugal.

Costa Branco P.J., Laboratorio de Mecatronica - CAUTL, Instituto Superior Técnico, CVRM - Centro de Valorização de Recursos Minerais, Av. Rovisco Pais, 1096 Lisboa Codex, Portugal.

Dubois D., Institut de Recherche en Informatique de Toulouse (IRIT), CNRS, Université Paul Sabatier, 118 Route de Narbonne, 31062 Toulouse Cedex, France.

Durão F.O., CVRMUTL - Departamento de Engenharia de Minas, Instituto Superior Técnico, CVRM - Centro de Valorização de Recursos Minerais, Av. Rovisco Pais, 1096 Lisboa Codex, Portugal.

Fargier H., Institut de Recherche en Informatique de Toulouse (IRIT), CNRS, Universite Paul Sabatier, 118 Route de Narbonne, 31062 Toulouse Cedex, France.

Gorrini V., IRIDIA, Université Libre de Bruxelles - CP 194/6, 50, ave. Franklin Roosevelt, 1050 Bruxelles, Belgium.

Harris C.J., University of Southampton, Department of Aeronautics and Astronautics, Highfield, Southampton SO17 1BJ, UK.

Holeňa M., University of Paderborn, Department of Computer Science, Cadlab, Fürstenallee 11, 33102 Paderborn, Germany.

John R.I., School of Computing Sciences, De Montfort University, The Gateway, Leicester LE1 9BH, UK.

Kovačević A., Clinical Hospital Centre, Spinčićeva 1, 58000 Split, Croatia.

Kovačević D., Maritime Faculty Dubrovnik, Department of Split University of Split, Zrinjsko-Frankopanska 38, 58000 Split, Croatia.

Martin T.P., University of Bristol, Department of Engineering Mathematics, Queen's Building, University Walk, Bristol, BS8 1TR, UK.

Pilsworth B., University of Bristol, Department of Engineering Mathematics, Queen's Building, University Walk, Bristol, BS8 1TR, UK.

Prade H., Institut de Recherche en Informatique de Toulouse (IRIT), CNRS, Université Paul Sabatier, 118 Route de Narbonne, 31062 Toulouse Cedex, France.

Rayner N.J.I., University of Southampton, Department of Aeronautics and Astronautics, Highfield, Southampton SO17 1BJ, UK.

Ulieru M., Brunel University, Department of Computer Science, Kingston Lane, Uxbridge, Middlesex UB8 3PH, UK.

Weber R., MIT, Korneliuscenter, Promenade 9, 52076 Aachen, Germany.

Zimmermann H., RWTH, Templergraben 64, 52056 Aachen, Germany.

1

Recent Developments in Fuzzy Logic and Intelligent Technologies

H.-J. Zimmermann

1.1 Introduction

1.1.1 Historical Sketch

Even though the first publication in the area of fuzzy set theory appeared in 1965 [4] the development of this theory for almost 20 years remained in the academic realm. The only exception was Japan, where Fuzzy Control, developed in Great Britain in the mid 1970s, was developed to commercial potential, which triggered similar developments in other parts of the world. Even now, after approximately 15,000 publications have appeared, most of them are theoretical in nature. The largest part of accessible (*i.e.* in English, French or German) publications on applications are still available in Germany. In Japan much stays hidden behind language barriers, and in the USA the number of applied publications is increasing slowly but steadily.

1.1.2 Application Areas

A variety of criteria can be used to classify applications of fuzzy set theory, which shall be called Fuzzy Technology (FT) in the following. From a methodological point of view, the following is probably the most consistent classification (excluding purely mathematical applications, *e.g.* fuzzy topology, etc.).

1. Algorithmic Applications, *e.g.*

>   Fuzzy Mathematical Programming
>
>   Fuzzy Planning Methods (CPM, Graphs)
>
>   Fuzzy Petri Nets
>
>   Fuzzy Clustering, *etc.*

2. Information Processing, *e.g.*

>   Fuzzy Data Bank System
>
>   Fuzzy Query Languages
>
>   Fuzzy Languages, *etc.*

3. Knowledge Based Applications, *e.g.*

>   Expert Systems
>
>   Fuzzy Control
>
>   Knowledge Based Diagnosis, *etc.*

4. Hybrid Application Areas

>   Fuzzy Data Analysis
>
>   Fuzzy Supervisory Control, *etc.*

These areas also correspond to the most important applications-oriented goals of FT:

1. **Relaxation** of dichotomous methods and models: *i.e.* powerful classical mathematical models are often not adequate modelling languages for non-dichotomous real problems of the 'more-or-less' type. Here FT is used to extend classical methods to increase their applicability. (Algorithmic applications.)

2. **Complexity Reduction** and improvement of the man-machine-interface. Primarily using the concept of a linguistic variable, the complexity of data masses is reduced to a degree which human beings can comprehend. Also, the link between formal or numerical 'languages' and linguistic human types of communication are established. (Information Processing, Data Mining.)

3. **Meaning Preserving Reasoning.** Also via linguistic variables and various inference methods of approximate reasoning, expert system technology is advanced from symbol processing systems to computer systems that are able to store, process and output linguistic expressions which combine words (as labels) with their meaning (defined as fuzzy sets. (Knowledge Based Systems.)

1.2 Recent Developments

1.2.1 Methodological Developments

Most of the methods in FT, such as Fuzzy Linear Programming, Fuzzy Clustering, Fuzzy Petri Nets, etc., were essentially developed in the 1970s or the early 1980s. They were refined until the beginning of the 1990s. Recent developments can primarily be found in the combination of, for example, FT with artificial neural nets, with evolutionary strategies and genetic algorithms and with chaos theory. In fuzzy control the merger of traditional control (PI, PID, etc.) with fuzzy controllers has also progressed. The motivation behind these developments is always to maintain the strengths of the approaches to be combined while avoiding their weakness.

1.2.2 New Application Areas

As already mentioned, the oldest area of FT applications is that of fuzzy control, particularly to consumer goods in Japan. This area is still flourishing by applying the original Mamdani-type controller or other types to consumer goods, or increasingly to industrial processes and products. In many cases, genetic algorithms and neural nets are added to either optimize parts of the controllers or to make them adaptive or learning. In addition to this original type of control, to an increasing degree supervisory control — coordinating many fuzzy controllers — is applied. This may then involve the use of Fuzzy Petri Nets, neural nets or other approaches. This type of application applies particularly to continuous production processes (such as in the chemical industry), but also traditional and nuclear power plants. Here FT is often combined with methods of multi-criteria analysis.

Very close to these applications lies the area of Fuzzy Data Analysis, which is quite young and rapidly increasing at present, and which uses fuzzy technology algorithmically as well as knowledge-based FT. The combination of FT with neural nets and genetic algorithms can also be found in this area. There are numerous other new or evolving application areas which cannot be discussed here.

1.2.3 Tools

FT is hardly ever applied to real problems by pencil and paper, but generally via computer technology. Until the end of the 1980s, for example, fuzzy controllers were programmed directly using higher computer languages such as Fortran, C or others. This is, of course, rather inefficient, particularly as long as no design methodology exists. At the end of the 1980s, one started to develop CASE tools, particularly for fuzzy control, which can be used to build fuzzy controllers much more efficiently. Nowadays at least 20 such tools exist. New developments are tools for Fuzzy Petri Nets and for the area of Fuzzy Data Analysis. The latter are more complicated than fuzzy control tools because they contain algorithmic methods (pre-processing methods and, for example, fuzzy clustering), knowledge-based modules (such as in fuzzy control), and neural nets, *e.g.* Fuzzy Kohonen nets, etc. The newest development in this direction is the use of genetic algorithms for clustering purposes. It should also be mentioned that quite a lot of fuzzy hardware has become available in the meantime, reaching from fuzzy ASICS (FASICS) to involved evaluation boards, *etc.*

1.3 Future Perspectives

One of the most important activities in the future will certainly be technology transfer, *i.e.* bringing suppliers of technology together with problem owners. In some areas of the world (*e.g.* Japan, Germany) activities in this direction have already processed quite far. In other territories they have just started or have not yet begun.

A lot of progress can be expected from growing cooperation and merging of FT with neural nets and genetic algorithms. This area is starting to be known as 'Computational Intelligence' on the three continents.

References

[1] Altrock v., C. (1993) *Fuzzy Logic*. Oldenbourg-Verlag.

[2] Bellman R.E. and Zadeh L.A. (1970) Decision-making in a Fuzzy Environment. *Management Science* 17: B141–B164.

[3] Bocklisch S.F. (1987) *Prozeßanalyse mit unscharfen Verfahren*. Berlin.

[4] Zadeh L.A. (1965) Fuzzy Sets. *Information and Control* 8: 338–353.

[5] Zimmermann H.-J. and Altrock v., C. (1994) *Fuzzy Logic*. Oldenbourg Verlag.

2

Handling Priority and Preference in Constraint Satisfaction Problems: A Possibility Theory-Based Approach

D. Dubois, H. Fargier, H. Prade

2.1 Introduction

Classical Constraint Satisfaction Problems (CSPs) only consider a set of hard constraints that every solution must satisfy. This rigid representation framework has several drawbacks. First, some problems are over-constrained and have no solutions. A relaxation of the less rigid or important constraints must be performed in order to obtain a solution. Discovering that a problem has no solution may be time-consuming, and devising an efficient constraint relaxation method is far from easy. Alternatively, other problems lead to a large set of equally possible solutions, although there often exist preferences among them which remain unexpressed. But a standard CSP procedure will pick a solution at random. Indeed, in practice, constraints are not always strict and it is desirable to extend the CSP framework in order to accommodate flexible constraints. Devising a framework for representing the flexibility of constraints will avoid artificially unfeasible problems (constraints being self-relaxable), and will avoid the random choice of solutions to loosely constrained problems. By flexible constraints, we mean either (i) *soft* constraints, which directly express preferences among solutions (*i.e.* this is a ranking of instantiations which are more or less acceptable for the satisfaction of a soft constraint), or (ii) *prioritized* constraints, that can be violated if they conflict with more prioritary constraints.

In soft constraints, the flexibility accounts for the possibility of going away from instantiations that satisfy the constraints ideally. Notice that the interest in soft constraints can be traced back to the early CSP literature; in 1975, Waltz [45] mentioned that he heuristically distinguished between 'likely' instantiations of a constraint and 'unlikely' ones which are only considered if necessary. Also in computer vision, in 1976 Rosenfeld *et al.* [39] modelled preference in the detection of convex objects for scene labelling problems, and proposed to use a fuzzy degree of constraint satisfaction. The idea of representing relative preferences by means of weights is also at work in the relaxation labelling process described in 1983 by Hummel and Zucker [31]. More recent works either propose to use fuzzy sets in modelling such constraints (Bowen *et al.* [3, 4], Guan and Friedrich [30], Freuder and Snow [27], Martin-Clouaire [36]) or to progressively relax the constraints when preferences are in conflict (Faltings *et al.* [19]).

In prioritized constraints the flexibility lies in the ability to discard constraints involved in inconsistencies, provided that they are not too important. Generally, a weight is associated with each constraint and the request is to minimize the greatest priority levels of the violated constraints (Descottes and Latombe [9]; Schiex [42]). More generally, Brewka *et al.* [5] and Borning *et al.* [2] identify different forms of constraint relaxation, viewing each constraint as a strict partial order on value assignment and weighting the importance of constraints; in particular, Brewka *et al.* [5] provide a formal semantics in relation to non-monotonic reasoning by means of maximal-consistent subsets of constraints.

Freuder [26], Freuder and Wallace [28] and Satoh [41] have devised theoretical foundations for the treatment of flexibility in CSPs. Satoh tries to apply results in non-monotonic reasoning based on circumscription to the handling of prioritized constraints so as to induce preference relations on the solution set. A similar point of view is adopted by Lang [33], where prioritized constraints are expressed in possibilistic logic (*i.e.* a logic with weighted formulas which has a non-monotonic behaviour in case of partial inconsistency). Taking a dual point of view, Freuder [26] regards a flexible problem as a collection of classical CSPs. A metric can then be defined that evaluates the distance between them. Then, the question is to 'find the solutions to the closest solvable problems'.

To take into account *both* types of flexibility, a generalization of the CSP framework has been proposed, the Fuzzy Constraint Satisfaction Problem (FCSP) framework (Dubois *et al.* [11], based on Zadeh [48]'s possibility theory (see [14] for an introduction)). The main point is that both types of flexible constraints are regarded as local criteria that rank-order (partial) instantiations and can be represented by means of fuzzy relations. In a FCSP, constraint satisfaction or violation are no longer an all-or-nothing notion: an instantiation is compatible with a flexible constraint to a degree (belonging to some totally ordered scale). The notion of consistency of a FCSP also becomes a matter of degree. The question is then to combine the satisfaction degrees of the fuzzy constraints in order to determine the total ordering induced over the potential solutions and to choose the best ones. Making a step further, we propose to use this framework also to handle more complex constraints, for example nested conditional constraints.

From an algorithmic point of view, the possibility of extending Waltz' algorithm to fuzzy constraints has been pointed out by Dubois and Prade [15] and by Yager [46]. As we will show, all the classical CSP algorithms (*e.g.* tree search, AC3, PC2) can easily be adapted to FCSPs. More generally, our framework reveals itself powerful enough to accommodate the definitions of local consistency of a problem (arc-consistency, 3-consistency, k-consistency). Interestingly enough, investigations by the second author (Fargier [21]) indicate that the theoretical results relating levels of local consistency of a CSP to its global consistency [25, 6] remain valid in FCSPs.

The next section deals with representation issues concerning flexible constraints. Fuzzy subsets on Cartesian products of domains, *i.e.* fuzzy relations, are used to model soft and/or prioritized constraints. An illustrative example is provided. The agreement of this representation with the preferential semantics of possibility theory is emphasized. Then the extension (resp. projection) of fuzzy constraints to larger (resp. smaller) Cartesian products of domains is recalled as well as the conjunctive or disjunctive combinations of fuzzy relations for representing compound constraints. Finally, this section discusses the modelling of more sophisticated constraints, namely prioritized constraints with safeguard (in order to guarantee the satisfaction of a weaker constraint in case of violation of the prioritized one). Then section 2.3 formally defines the FCSP framework and compares it to other approaches to flexibility in CSP. This section then presents the essentials of a Branch and Bound algorithm performing the search for the best solutions. Non-monotonic aspects of FCSPs are also outlined. Different notions of local consistency (arc-consistency, k-consistency) of a FCSP are defined in section 2.4; the complexity of extensions of filtering algorithms (*e.g.* AC3) to the fuzzy set framework is also discussed.

2.2   Representing Flexible Constraints

A hard constraint C relating a set of decision variables $\{x_1, \ldots, x_n\}$ ranging on respective domains $D_1, \ldots, D_n$ is classically described by an associated relation R : R is the crisp subset of $D_1 \times \cdots \times D_n$ that specifies the tuples $d = (d_1, \ldots, d_n)$ of values which are compatible with C. The set $\{x_1, \ldots, x_n\}$ of variables related by R will be denoted by $V(R)$.

2.2.1   Fuzzy Model of a Soft Constraint

A soft constraint C will be described by means of an associated *fuzzy* relation R [47], *i.e.* the fuzzy subset of $D_1 \times \cdots \times D_n$ of values that more or less satisfy C. R is defined by a membership function μ_R which associates a level of satisfaction $\mu_R(d_1, \ldots, d_n)$ in a totally ordered set L (with top denoted 1 and bottom denoted 0) to each tuple $(d_1, \ldots, d_n) \in D = D_1 \times \cdots \times D_n$. This membership grade indicates to what extent $d = (d_1, \ldots, d_n)$ is compatible with (or satisfies) C. Thus, the notion of constraint satisfaction becomes a matter of degree:

$$\mu_R(d_1, \ldots, d_n) = 1 \quad \text{means } (d_1, \ldots, d_n) \quad \text{totally satisfies } C;$$
$$\mu_R(d_1, \ldots, d_n) = 0 \quad \text{means } (d_1, \ldots, d_n) \quad \text{totally violates } C;$$
$$0 < \mu_R(d_1, \ldots, d_n) < 1 \quad \text{means } (d_i, \ldots, d_n) \quad \text{partially satisfies } C.$$

Hard constraints are particular cases of soft constraints, since they involve levels 0 and 1 only. A soft constraint involving preferences between values is regarded as a local criterion ordering the instantiations of C, preference levels being represented in the scale $L : \mu_R(d_1, \ldots, d_n) > \mu_R(d'_1, \ldots, d'_n)$ means that the first instantiation is preferred to the second one. Interpreting the preference degrees as membership degrees leads to the representation of a soft constraint by a fuzzy relation.

The assumption of a totally ordered satisfaction scale underlying the above setting may be questioned. The very use of a satisfaction scale instead of just an ordering relation is crucial when it comes to the aggregation of local satisfaction levels. Indeed, due to the famous Arrow theorem [38], it is very difficult to merge several ordering relations that are not commensurate. The satisfaction scale need not be totally ordered, stricly speaking, since a complete lattice will do as well. In the following we assume that L is a totally ordered set, *i.e.* a chain. But the scale of membership grades need not be numerical, as pointed out years ago [29]. A qualitative scale makes sense on finite domains. However on continuous domains, as in the case of temporal constraints with continuous time, it is much more natural and simple to assume that the satisfaction scale is the unit interval; then levels of satisfaction reflect distances to ideal values in the domain.

2.2.2   Fuzzy Model of a Prioritized Constraint

Fuzzy relations also offer a suitable formalism for the expression of prioritized constraints. When it is possible to *a priori* exhibit a total preorder over the respective priorities of the constraints, these priorities will be represented by levels in another scale V : a priority degree $Pr(C)$ is attached to each constraint C and indicates to what extent it is imperative that C be satisfied. First consider the case of hard constraints. $Pr(C) = 1$ means that C is an absolutely imperative constraint, while $Pr(C) = 0$ indicates that it is completely possible to violate C (C has no incidence in the problem). Given two constraints C and C', $Pr(C) > Pr(C')$ means that the satisfaction of C is more necessary than the satisfaction of C'. If C and C' cannot be satisfied simultaneously, solutions compatible with C will be preferred to solutions compatible with C'.

In fact, the scale V can be interpreted as a 'violation scale': the greater $Pr(C)$, the less it is possible to violate C. This remark leads us to relate the satisfaction scale L to the violation scale V, considering that there exists an order-reversing bijection from V to L such that $L = c(V) = \{c(v), v \in V\}$: $c(0)$ and $c(1)$ are respectively the top element and the bottom element of L, and $v \geq v'$ in V implies $c(v) \leq c(v')$ in L. This is one of the basic modelling assumptions in this chapter: the c-complement of the level of priority of a constraint is interpreted as the extent to which the constraint can be violated, using the reversed priority scale $L = c(V)$ as a satisfaction scale; L is nothing but V put upside down. Since $Pr(C)$ represents to what extent it is necessary to satisfy C, $c(Pr(C))$ indicates to what extent it is possible to violate C, i.e. to satisfy its negation. In other words, the constraint C is considered as satisfied at least to degree $c(Pr(C))$ whatever the considered solution, whether it satisfies C or not. More precisely, the prioritized constraint $(C, Pr(C))$ is considered as totally satisfied by a tuple if C is satisfied, and satisfied to degree $c(Pr(C))$ if the tuple violates C. Hence $c(V)$ can be identified to a satisfaction scale as in the previous section, and a prioritized constraint C may be represented by the fuzzy relation (see Figure 2.1).

$$\mu_R(d_1, \ldots, d_n) = c(0) = 1 \quad \text{if} \quad (d_1, \ldots, d_n) \text{ satisfies } C;$$
$$\mu_R(d_1, \ldots, d_n) = c(Pr(C)) = 1 \quad \text{if} \quad (d_1, \ldots, d_n) \text{ violates } C.$$

Note that when $Pr(C) = 1$, the characteristic function of C is recovered, while when $Pr(C) = 0$ the constraint C degenerates into the whole domain D.

Conversely, a soft constraint C where preferences are described in terms of a finite number of satisfaction degrees $0 = \alpha_0 < \alpha_1 < \cdots < \alpha_p < 1$ in a scale L, can be represented by a finite set of prioritized constraints $\{C^j, 0 \leq j < p\}$ using the scale L put upside down as a priority scale, via an order-reversing map c:

$$Pr(C^j) = c(\alpha_j) \quad \text{defining } R^j \quad = \quad \{(d_1, \ldots, d_n), \mu_R(d_1, \ldots, d_n) \geq \alpha_{j+1}\},$$
$$j = 0, p - 1.$$

If, moreover, it is assumed that c is involutive, that is $c(c(\alpha)) = \alpha$ (this hypothesis is made throughout the whole chapter), then it is straightforward to reconstruct the soft constraint C by means of the set of prioritized constraints $\{(C^j, Pr(C^j)), 0 \leq j < p\}$ as shown in Figure 2.2, where:

$$\mu_R(d) = \min_j \max(c(Pr(C^j)), \mu_{R^j}(d)) \text{ for every tuple } d = (d_1, \ldots, d_n).$$
$$\tag{2.1}$$

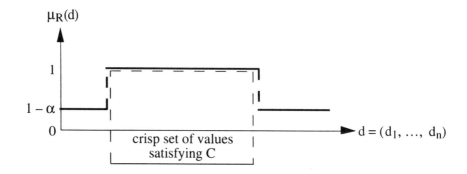

Figure 2.1: A hard (or crisp) constraint C with priority $Pr(C) = \alpha$ when $c(x) = 1 - x$.

Finally, a prioritized soft constraint C corresponds to the following fuzzy relation:

$$\mu_{R'}(d_1, \ldots, d_n) = \max(c(Pr(C)), \mu_R(d_1, \ldots, d_n)), \qquad (2.2)$$

where R is the fuzzy relation describing the preferences of C only. Viewing the soft constraint expressed by R as a family of nested prioritized constraints, the global priority $Pr(C)$ attached to the soft constraint C means that we forget the priorities higher than $Pr(C)$ in the expression of R, since

$$\max(c(Pr(C)), \mu_R(d)) = \max(c(Pr(C)), \min_j \max(c(Pr(C^j)), \mu_{R^j}(d)))$$

$$= \min_j \max(c(\min(Pr(C), Pr(C^j))), \mu_{R^j}(d)).$$

To conclude with representation issues, prioritized and soft constraints can be cast in a unique setting that we call 'flexible constraints', modelled by fuzzy sets, where flexibility means the capability of self-relaxation. This capability is locally embedded in the description of the constraint, thus avoiding the necessity of a specific constraint relaxation procedure to be triggered when a set of constraints is found inconsistent. This unification presupposes a strong link between levels of constraint satisfaction and levels of constraint priority, using a single ordered scale L for both priority and satisfaction and an order-reversing map c that changes one notion into the other. For simplicity, we

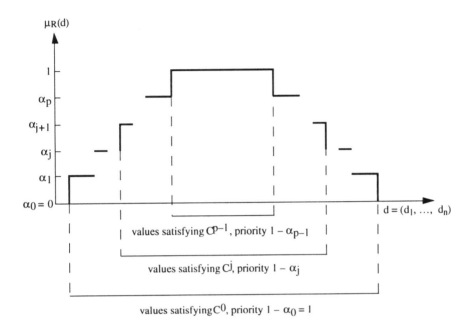

Figure 2.2: Decomposition of a soft constraint into a family of prioritized constraints when $c(x) = 1 - x$.

sometimes use $L = [0, 1]$ and $c(x) = 1 - x$ in the following. However, all results to be presented remain valid on a qualitative scale.

2.2.3   Example

A course must involve seven sessions, namely x lectures, y exercise sessions and z training sessions ($C1$). There must be about two training sessions ($C2$), *i.e.* ideally 2, possibly 1 or 3. Dr. B, who gives the exercise part of the course, wants to manage 3 or 4 sessions ($C3$). Prof. A, who gives the lectures, wants to give about 4 lectures ($C4$), *i.e.* ideally 4 lectures, possibly 3 or 5). The request of Dr. B is less important than the one of Prof. A and is itself less important than the imperative constraints $C1$ and $C2$. In this example, flexibility is modelled using a five level scale $L = (\alpha_0 = 0 < \alpha_1 = c(\alpha_3) < \alpha_2 = c(\alpha_2) < \alpha_3 = c(\alpha_1) < \alpha_4 = 1)$, where c is the order-reversing operation. The priorities of $C3$ and $C4$ are, respectively, α_2 and α_3 ($\alpha_2 < \alpha_3$). The domain of variables x, y and z is the set $\{0,1,2,3,4,5,6,7\}$. The following

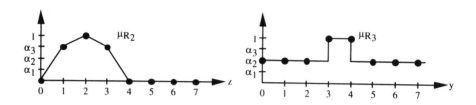

Figure 2.3: Modelling of C_2 (a) and C_3 (b) by means of fuzzy unary restrictions.

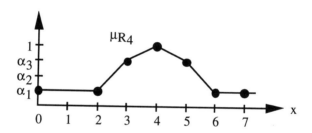

Figure 2.4: Modelling of C_4 by means of a fuzzy unary restriction.

model can be used:

C1: classical hard constraint

$\mu_{R_1}(x, y, z) = 1$ if $x + y + z = 7$;

$\mu_{R_1}(x, y, z) = 0$ otherwise.

C2: soft constraint (see Figure 2.3a)

$\mu_{R_2}(z) = 1$ if $z = 2$;

$\mu_{R_2}(x) = \alpha_3$ if $z = 1$ or $z = 3$;

$\mu_{R_2}(z) = 0$ otherwise.

C3: prioritized constraint $Pr(C_3) = \alpha_2$ (see Figure 2.3b)

$\mu_{R_3}(y) = 1$ if $y = 3$ or $y = 4$;

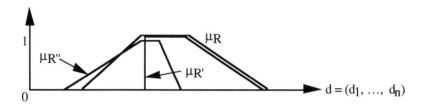

Figure 2.5: $R' \subseteq R$ and $R'' \not\subseteq R$.

$\mu_{R_3}(y) = c(\alpha_2) = \alpha_2$ otherwise.

C4: soft and prioritized constraint $Pr(C4) = \alpha_3$ (see Figure 2.4)

$\quad \mu_{R_4}(x) = 1$ if $x = 4$;

$\quad \mu_{R_4}(x) = \max(\alpha_3, c(\alpha_3)) = \alpha_3$ if $x = 3$ or $x = 5$;

$\quad \mu_{R_4}(x) = c(\alpha_3) = \alpha_1$ otherwise.

2.2.4   Operations on Fuzzy Relations

Flexible constraints are modelled by qualitative fuzzy relations. The usual operations on crisp relations can be easily generalized to fuzzy relations [47]. To do so, we exploit the fact that, being totally ordered, the satisfaction scale L is a complete distributive lattice, where the minimum and the maximum of two elements make sense. The following definitions extend classical set-theoretic notions used in constraint-directed problem-solving:

- A fuzzy relation R' is said to be *included* into R if and only if (see Figure 2.5):

$$\forall (d_1, \ldots, d_n) \in D_1 \times \cdots \times D_n, \mu_{R'}(d_1, \ldots, d_n) \leq \mu_R(d_1, \ldots, d_n).$$

  This definition is a generalization of the classical set inclusion. In terms of constraints, C' is tighter that C and C is a relaxation (or a weakening) of C'.

- The *projection* of a fuzzy relation R on $\{x_{k_1}, \ldots, x_{k_{n_k}}\} \subseteq V(R)$ is a fuzzy relation $R^{\downarrow \{x_{k_1}, \ldots, x_{k_{n_k}}\}}$ on $\{x_{k_1}, \ldots, x_{k_{n_k}}\}$ such that:

$$\mu_{R^{\downarrow \{x_{k_1}, \ldots, x_{k_{n_k}}\}}}(d_{k_1}, \ldots, d_{k_{n_k}}) = \sup_{\{d/d^{\downarrow \{x_{k_1}, \ldots, x_{k_{n_k}}\}} = (d_{k_1}, \ldots, d_{k_{n_k}})\}} \mu_R(d)$$

  where $d^{\downarrow \{x_{k_1}, \ldots, x_{k_{n_k}}\}}$ denotes the classical restriction of $d = (d_1, \ldots, d_n)$ to $\{x_{k_1}, \ldots, x_{k_{n_k}}\}$. This definition is a generalization of the projection

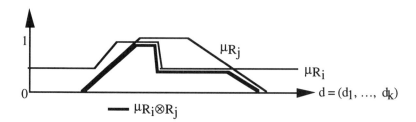

Figure 2.6: Conjunctive combination of two fuzzy relations R_i and R_j.

of ordinary relations. $\mu_{R^{\uparrow\{x_{k_1},\dots,x_{k_{n_k}}\}}}(d_{k_1},\dots,d_{k_{n_k}})$ estimates to what level of satisfaction the instantiation $(d_{k_1},\dots,d_{k_{n_k}})$ can be extended to an instantiation that satisfies C.

- The *cylindrical extension* of a fuzzy relation R to $\{x_{k_1},\dots,x_{k_{n_k}}\} \supseteq V(R)$ is a fuzzy relation $R^{\uparrow\{x_{k_1},\dots,x_{k_{n_k}}\}}$ on $\{x_{k_1},\dots,x_{k_{n_k}}\}$ such that:

$$\mu_{R^{\uparrow\{x_{k_1},\dots,x_{k_{n_k}}\}}}(d_{k_1},\dots,d_{k_{n_k}}) = \mu_R((d_{k_1},\dots,d_{k_{n_k}})^{\downarrow V(R)}).$$

This definition is a generalization of the cylindrical extension of ordinary relations. $\mu_{R^{\uparrow\{x_{k_1},\dots,x_{k_{n_k}}\}}}(d_{k_1},\dots,d_{k_{n_k}})$ estimates to what extent the instantiation $(d_{k_1},\dots,d_{k_{n_k}})$ satisfies C.

- The *conjunctive combination* (or join) of two fuzzy relations R_i and R_j is a fuzzy relation $R_i \otimes R_j$ over $V(R_i) \cup V(R_j) = \{x_1,\dots,x_k\}$ such that (see Figure 2.6):

$$\mu_{R_i \otimes R_j}(d_1,\dots,d_k) = \min(\mu_{R_i}((d_1,\dots,d_k)^{\downarrow V(R_i)}),$$
$$\mu_{R_j}((d_1,\dots,d_k)^{\downarrow V(R_j)})).$$

$\mu_{R_i \otimes R_j}(d_1,\dots,d_k)$ estimates to what extent $(d_1,\dots,d_k)$ satisfies both C_i and C_j. When $V(R_i) = V(R_j), \otimes$ is a generalization of classical set intersection. All properties of the standard intersection (associativity, commutativity, etc.) hold as long as negation is not involved; in particular, there holds $(R_i \otimes R_j)^{\downarrow V(R_i)} \subseteq R_i$ and $(R_i \otimes R_i) = R_i$.

Note that the use of the combination rule, allowed by the presence of a unique satisfaction scale, underlies an assumption of *commensurability* between satisfaction levels pertaining to different constraints: the user who specifies the constraints must describe them by means of this *unique scale L* (or by means of the dual scale L^T). For instance, in the example of section 2.2.3, the satisfaction level α_3 of C_4 for $x \in \{3,5\}$ is assumed

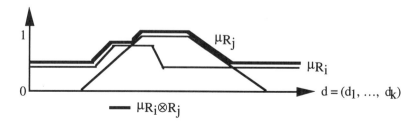

Figure 2.7: Disjunctive combination of two fuzzy relations R_i and R_j.

to be equal to the satisfaction level for $z \in \{1, 3\}$ and $\alpha_1 < c(Pr(C_3)) < c(Pr(C_4))$. Although natural and often implicit, this assumption must be emphasized.

- The *disjunctive combination* of two fuzzy relations R_i and R_j is a fuzzy relation $R_i \oplus R_j$ over $V(R_i) \cup V(R_j) = \{x_1, \ldots, x_k\}$ such that (see Figure 2.7):

$$\mu_{R_i \oplus R_j}(d_1, \ldots, d_k) = \max(\mu_{R_i}((d_1, \ldots, d_k)^{\downarrow V(R_i)}),$$
$$\mu_{R_j}((d_1, \ldots, d_k)^{\downarrow V(R_j)})).$$

$\mu_{R_i \oplus R_j}(d_1, \ldots, d_k)$ estimates to what extent $(d_1, \ldots, d_k)$ satisfies either C_i or C_j. When $V(R_i) = V(R_j)$, $\oplus$ is a generalization of classical set union. All properties of set union (associativity, commutativity, distributivity over intersection, etc.) hold, if negation is not involved.

2.2.5 Prioritized Constraints with Safeguard

The framework of fuzzy constraints offers a convenient tool for representing more sophisticated constraints than the previously encountered ones, for instance prioritized constraints with safeguard, as well as nested conditional constraints, as we are going to see. First one may like to express that a constraint C, even with a rather low priority $Pr(C) = \alpha$, cannot ever be completely violated, in the sense that if C is violated, at least a more permissive, minimal, constraint C' is still satisfied. Let R and R' be the fuzzy relations associated with C and C', respectively, with $R \subseteq R'$ (C' is more permissive than C, *i.e.* C' is a relaxation of C). The whole constraint C^* corresponding to the pair (C, C') can be viewed as the conjunction of a prioritized constraint (C) and a weaker but imperative, possibly soft, constraint (C'). This conjunction is represented by the fuzzy relation R^*, pictured in Figure 2.8, and

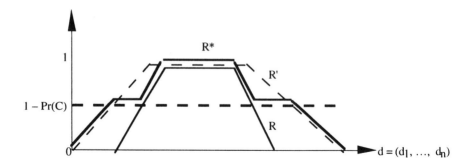

Figure 2.8: Representation of a prioritized fuzzy constraint with safeguard.

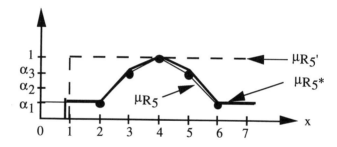

Figure 2.9: Modelling of C_5 by fuzzy constraint with safeguard constraint.

expressed by:

$$\forall d \in D_1 \times \cdots \times D_n, \mu_{R^\bullet}(d) = \min(\max(\mu_R(d), c(\alpha)), \mu_{R'}(d)).$$

$$(2.3)$$

This is a particular case of the decomposition of a soft constraint into prioritized ones when C and C' are hard. Indeed, such constraints express both a requirement with priority α less than 1 and a weaker requirement with priority 1 and R^* is of the form (2.1):

$$\mu_{R^\bullet}(d) = \min(\max(\mu_R(d), c(\alpha)), \max(\mu_{R'}(d), c(Pr(C')))) \text{ with } Pr(C') = 1.$$

See [18] for the use of such constraints in fuzzy database querying systems. Interestingly enough, R^* can be decomposed either as a disjunction or as a conjunction of two fuzzy relations, depending on which fuzzy relation, R or

R', the priority weight is combined with. Indeed,

$$
\begin{aligned}
\mu_{R^*}(d) &= \min(\max(\mu_R(d), c(\alpha)), \mu_{R'}(d)) \\
&= \min(\max(\mu_R(d), c(\alpha)), \max(\mu_R(d), \mu_{R'}(d))) \text{ since } R \subseteq R' \\
&= \max(\mu_R(d), \min(c(\alpha), \mu_{R'}(d))).
\end{aligned}
$$

It expresses that satisfying a constraint with safeguard corresponds to either satisfying its stronger form C, or its weaker form C', the satisfaction degree being upper-bounded in this second case by $c(\alpha)$.

For instance, a flexible C_5 constraint prescribing: 'Prof. A wants to give about four lectures; anyway, he will never accept to give no lecture' is represented by the fuzzy relation R_5^* pictured in Figure 2.9.

2.3 Stating and Solving Fuzzy Constraint Satisfaction Problems

2.3.1 Definition

A Fuzzy Constraint Satisfaction Problem (FCSP) P involves a set of n decision variables $X = \{x_1, \ldots, x_n\}$ each ranging on its respective domain $D_1, \ldots, D_n$ and a set of m fuzzy relations $R = \{R_1, \ldots, R_m\}$ representing a set $C = \{C_1, \ldots, C_m\}$ of hard, soft or prioritized constraints (domains are assumed to be discrete in the following). A unary relation R_j of R is supposed to be associated to each variable x_j. It represents the values which are *a priori* more or less feasible (*i.e.* here, preferred) for x_j (by default, $R_j = D_j$). If all the constraints are unary or binary, the FCSP is called a fuzzy constraint network.

Classically, an instantiation of $\{x_{k_1}, \ldots, x_{k_{n_k}}\} \subseteq X$ is locally consistent if it satisfies all the constraints in the subnetwork restricted to $\{x_{k_1}, \ldots, x_{k_{n_k}}\}$. Like constraint satisfaction, the notion of consistency is now a matter of degree. The definition of the conjunctive combination states that

$$
\mu_{[R_{i_1} \oplus \cdots \oplus R_{i_p}]}(d_{k_1}, \ldots, d_{k_{n_k}})
$$

estimates to what extent $(d_{k_1}, \ldots, d_{k_{n_k}})$ satisfies all the constraints $C_{i_1}, \ldots, C_{i_p}$. Hence, the degree of local consistency of $(d_{k_1}, \ldots, d_{k_{n_k}})$ is defined by:

$$
\begin{aligned}
&\text{Cons}(d_{k_1}, \ldots, d_{k_{n_k}}) \\
&= \mu_{[\oplus_{\{R_i \in R / V(R_i) \subseteq \{x_{k_1}, \ldots, x_{k_{n_k}}\}\}} R_i]}(d_{k_1}, \ldots, d_{k_{n_k}}) \\
&= \min_{\{R_i \in R / V(R_i) \subseteq \{x_{k_1}, \ldots, x_{k_{n_k}}\}\}} (\mu_{R_i}((d_{k_1}, \ldots, d_{k_{n_k}})^{\downarrow V(R_i)})).
\end{aligned}
$$

It should be noticed that:

$$
\forall Y \subseteq \{x_{k_1}, \ldots, x_{k_{n_k}}\}, \text{Cons}((d_{k_1}, \ldots, d_{k_{n_k}})^{\downarrow Y}) \geq \text{Cons}(d_{k_1}, \ldots, d_{k_{n_k}}).
$$

Considering a complete instantiation of $X, \mu_{[R_1 \otimes \cdots \otimes R_m]}(d_1, \ldots, d_n)$ is the satisfaction degree of all the constraints by $(d_1, \ldots, d_n)$, *i.e.* the satisfaction degree of problem P by $(d_1, \ldots, d_n)$. It is the membership degree of $(d_1, \ldots, d_n)$ to the fuzzy set $\rho = R_1 \otimes \cdots \otimes R_m$ which is nothing but the (fuzzy) set of solutions of P. As for classical CSPs, solutions are consistent instantiations of X : $\text{Cons}(d_1, \ldots, d_n) = \mu\rho(d_1, \ldots, d_n) > 0$, *i.e.* solutions that are not totally unfeasible.

These degrees discriminate among the potential solutions since they induce a total preorder over the instantiations; this preorder does not depend on whether L is a numerical scale or not. In other words, the FCSP approach to flexibility is more qualitative than quantitative. Actually, solving a classical CSP means separating the set of all instantiations into two classes: the instantiations which are solutions to the problem, and those which are not. Introducing flexibility just refines this order.

It should be noticed that the best instantiations of X may get a satisfaction degree lower than 1 if some constraints are conflicting: the FCSP approach can handle partially inconsistent problems. The consistency degree of the FCSP is the satisfaction degree of the best instantiations:

$$
\begin{aligned}
\text{Cons}(P) \quad &= \quad \sup_{\{(d_1, \ldots, d_n) \in D_1 \times \cdots \times D_n\}} \mu\rho(d_1, \ldots, d_n) \\
&= \quad \sup_{\{(d_1, \ldots, d_n) \in D_1 \times \cdots \times D_n\}} [\min_{\{R_i \in R\}} \mu_{R_i}((d_1, \ldots, d_n)^{\downarrow V(R_i)})].
\end{aligned}
$$

The best solutions of P are those which satisfy the global problem to the maximal degree $\mu_{R_1 \otimes \cdots \otimes R_m}(d_1, \ldots, d_n) = \mu\rho(d_1, \ldots, d_n)$ ($= \text{Cons}(P)$), *i.e.* those which maximize the satisfaction level of the least satisfied constraint. If there are some instantiations which perfectly satisfy all the constraints ($\text{Cons}(P) = 1$), they are the best solutions. Otherwise, an implicit relaxation of flexible constraints is performed, achieving a trade-off between antagonistic constraints in the spirit of [9]: a solution will be found as long as the problem is not totally inconsistent.

Our example of section 2.2.3 is partially inconsistent. The best solution is $(x = 3, y = 3, z = 1)$ and the consistency degree is $\alpha_3 < 1$: the constraint over the number of training sessions and Prof. A's constraint are slightly relaxed according to their flexibility. The other potential solutions (*e.g.* $(x = 4, y = 1, z = 2)$ or $(x = 2, y = 3, z = 2)$) are less consistent (their respective satisfaction degrees are α_2 and α_1).

2.3.2 Discussion

The FCSP approach is in accordance with Freuder's view of constraint relaxation by partial satisfaction [26]. Indeed, a FCSP involving p different satisfaction levels is equivalent to p CSPs: for each level $\alpha_j > 0, \alpha_j \in L$, a CSP P^{α_j} is formed by the set of hard constraints $C_i^{\alpha_j}$ containing the tuples that satisfy C_i to a degree greater than or equal to α_j. Considering that a weight is associated to each possible relaxation of each constraint C_i, the metric associated to this space is defined by the maximum among the weights of the relaxations performed. The set of best solutions to the flexible problem is the set of solutions of the consistent P^{α_j} of highest α_j (the closest solvable problem using Freuder's terminology).

The FCSP approach is different from probabilistic or cost-based approaches, where the best solutions are those satisfying the maximal number of constraints [28] or those for which the sum of satisfaction degrees is maximal [40]. These additive approaches allow for the violation of a constraint to be counterbalanced by the satisfaction of other constraints. The word 'constraint' is then hardly justified. In a FCSP, as soon as an instantiation violates a hard constraint it is totally inconsistent:$\mu_{R_1 \otimes \cdots \otimes R_m}(d_1, \ldots, d_n) = 0$. Thus, we are in accordance with the principle of constraint satisfaction: no constraint can be violated except according to its relaxation capacities, which are expressed by the FCSP formalism. Additive satisfaction pooling methods also presuppose that constraints are independent, or at least not redundant. This ideal is difficult to achieve and looks contradictory with the purpose of constraint propagation, which is to produce redundant constraints. Note that the two methods of aggregation of satisfaction levels correspond to the two basic approaches to the definition of social welfare in utility theory (*e.g.* [38]): utilitarianism which maximizes the sum of the individual utilities, and egalitarianism which maximizes the minimal individual utility. Only the latter is compatible with the usual treatment of constraints.

Although in accordance with ours, Satoh [41]'s approach differs in the way in which priorities between constraints are expressed. Indeed, Satoh uses second-order logic to describe priorities. Moreover, the ordering of solutions depends on how many constraints are satisfied. In our approach, solutions which satisfy a FCSP to the same degree are not discriminated, even if some of them satisfy more constraints. In other terms, the best solutions in the sense of Satoh are among the best according to the FCSP definition. However, a so-called lexicographic ordering may be used in FCSP, if needed, to discriminate solutions sharing the same global satisfaction degree, as for instance in [9, 27]. This mode of aggregation is also known in the social welfare literature under the name 'leximin aggregation' (see [38]). The definition of the leximin ordering of two vectors $v_1 = (\mu_1, \ldots, \mu_n)$ and $v_2 = (\lambda_1, \ldots, \lambda_n)$ in L^n is as follows:

1. rearrange the vectors in increasing order, say $\mu_{i_1} \leq \mu_{i_2} \leq \cdots \leq \mu_{i_n}$ and $\lambda_{j_1} \leq \lambda_{j_2} \leq \cdots \leq \lambda_{j_n}$;

2. perform a lexicographical comparison starting from the first component, *i.e.*

$$v_2 > v_1 \Leftrightarrow \exists k \leq n \text{ such that } \forall m < k$$
$$\lambda_{j_m} = \mu_{i_m} \text{ and } \lambda_{j_k} > \mu_{i_k}.$$

In the example, the instantiation ($z = 2, y = 1, x = 4$) which satisfies C_1, C_2, C_3 and C_4 to degrees $(1, 1, \alpha_2, 1)$ is considered in a FCSP as equally good as $(z = 1, y = 1, x = 5)$ which satisfies the constraints to degrees $(1, \alpha_3, \alpha_2, \alpha_3)$. The lexicographic ordering, which is a refinement of the min-induced ordering, will prefer the first instantiation to the second one. Note that if $L = \{0, 1\}$, *i.e.* if the FCSP is a classical CSP, the solutions which are the best according to the lexicographic ordering are those satisfying the maximal number of constraints, as in Freuder's view of partial constraint satisfaction [28]. In other words, the lexicographic ordering in a FCSP, which is more precisely studied in [22, 12], is a generalization of Freuder's ordering in a classical CSP.

As a general model based on possibility theory, the FCSP approach generalizes the frameworks that model softness by means of fuzzy sets [3, 27, 36] as well as those dealing with constraint priorities by searching to minimize the priority of the violated constraints [9, 27, 2, 42]. More precisely, some of them use an inclusion-based refinement of the min-induced ordering [2], or a lexicographic refinement [9, 27] which is itself a refinement of the inclusion-based ordering. See [22, 12] for a discussion on the selection of preferred solutions in FCSP by means of these three criteria.

2.3.3   A Generic Solving Method for FCSPs

Finding a solution to a classical CSP is a NP-complete task. Hence, finding the best solution of FCSP is at least NP-hard. In fact, it reduces to a sup/min optimization formulation:

$$\sup_{\{(d_1,\ldots,d_n)\in D_1\times\cdots\times D_n\}} [\min_{\{R_i\in\{R_1,\ldots,R_m\}\}} (\mu_{R_i}((d_1,\ldots,d_n)^{\downarrow V(R_i)}))].$$

This kind of problem can be solved using classical Branch and Bound algorithms [26, 33, 42], such as Depth-First Branch and Bound. It is a natural extension of backtracking, the standard approach to CSPs. Using such a classical tree search algorithm, variables are instantiated in a predetermined sequence, say $(x_1, \ldots, x_n)$. The root of the tree is the empty assignment. Intermediate nodes $(d_1, \ldots, d_k)$ denote partial instantiations and leaves are complete instantiations of $(x_1, \ldots, x_n)$. For each leaf $(d_1, \ldots, d_n)$ in the tree, we may compute $\mu_\rho(d_1, \ldots, d_n)$. The leaves that maximize μ_ρ are searched for via a depth-first exploration of the tree.

The use of fuzzy constraints makes it possible to prune each branch that necessarily leads to suboptimal leaves that can be proved worse than the best of the already evaluated solutions. In other words, it is useless to extend intermediary nodes $(d_1, \ldots, d_k)$ such that $\mu_{[\rho\downarrow\{x_1,\ldots,x_k\}]}(d_1, \ldots, d_k) \leq b_{\inf}$, $b_{\inf}$ being a lower bound of $\mathrm{Cons}(P)$. The calculation of $\mu_{[\rho\downarrow\{x_1,\ldots,x_k\}]}(d_1, \ldots, d_k)$ requires the extension of$(d_1, \ldots, d_k)$ into a complete instantiation, but the definition of local consistency provides an upper bound for it. Indeed:

$$\mathrm{Cons}(d_1, \ldots, d_k)$$

$$= \mu_{[\otimes_{\{R_i \in R / V(R_i) \subseteq \{x_1, \ldots, x_k\}\}} R_i]}(d_1, \ldots, d_k)$$

$$= \min_{\{R_i \in R / V(R_i) \subseteq \{x_1, \ldots, x_k\}\}} (\mu_{R_i}((d_1, \ldots, d_k)^{\downarrow V(R_i)}))$$

$$\geq \min_{\{R_i \in R\}} (\mu_{R_i}((d_1, \ldots, d_k)^{\downarrow V(R_i)})) = \mu_{[\rho^{\downarrow \{x_1, \ldots, x_k\}}]}(d_1, \ldots, d_k).$$

Hence $\mathrm{Cons}\,(d_1, \ldots, d_k) \geq \mu_{[\rho^{\downarrow \{x_1, \ldots, x_k\}}]}(d_1, \ldots, d_k)$.

This bound decreases when extending the nodes of the search tree and becomes exact for the leaves. Moreover, it may be incrementally computed as the tree is explored downward:

$$\mathrm{Cons}(d_1, \ldots, d_{k+1}) = \min(\mathrm{Cons}(d_1, \ldots, d_k),$$

$$\min_{\{R_i \in R, x_{k+1} \in V(R_i) \text{ and } V(R_i) \subseteq \{x_1, \ldots, x_{k+1}\}\}} \mu_{R_i}((d_1, \ldots, d_{k+1})^{\downarrow V(R_i)}))).$$

Like the incremental computation of consistency in classical CSPs, the incremental computation of $\mathrm{Cons}(d_1, \ldots, d_{k+1})$ considers each constraint only once.

Hence, the search starts with a lower bound b_{inf} (for pruning) and an upper bound b_{sup} of $\mathrm{Cons}(P)$; b_{inf} and b_{sup} may respectively be initialized to 0 and 1, or to better lower and upper bounds of $\mathrm{Cons}(P)$ if available. The consistency of the root is taken as b_{sup}. At each step, the current partial instantiation $(d_1, \ldots, d_k)$ is tentatively extended to variable x_{k+1}. If there is a value d_{k+1} such that $\mathrm{Cons}(d_1, \ldots, d_{k+1}) > b_{inf}, d_{k+1}$ is assigned to x_{k+1}. If no value consistent enough can be found for x_{k+1}, the algorithm backtracks to the most recent variable assignment. When a solution $(d_1, \ldots, d_n)$ is reached whose consistency is greater than b_{inf}, it is thus the best current solution; b_{inf} is updated to $\mathrm{Cons}(d_1, \ldots, d_n)$ since it is a better lower bound of $\mathrm{Cons}(P)$. If $\mathrm{Cons}(d_1, \ldots, d_n) < b_{sup}$, the algorithm backtracks to find a solution better than the current one. It should be noticed that partial instantiations $(d_1, \ldots, d_k)$ which have been extended to a solution $(d_1, \ldots, d_n)$ whose consistency is equal to $\mathrm{Cons}(d_1, \ldots, d_k)$ do not have better extensions; hence, these extensions do not have to be explored.

Figure 2.10 shows a search tree corresponding to the example of section 2.2.3.

Circumstances may impose resource bounds. In particular, real time processing may require immediate answers that can be refined later if time allows. The Depth-First Branch and Bound process is well suited to provide resource-bounded solutions. We can simply report the best instantiation available when, for example, a time limit is exceeded. In our example, the discovery of the best solution requests 10 nodes and 37 checks of consistency (*i.e.* computations of the satisfaction degree of a constraint): the first solution (consistency: α_1) is reached after three node extensions and 10 consistency checks and the best solution (consistency: α_3) is encountered after seven node extensions and

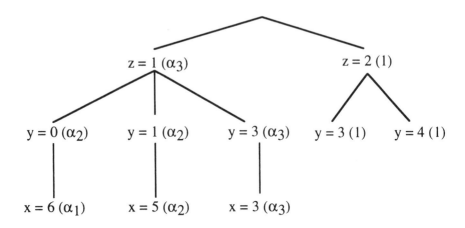

Figure 2.10: A search tree for the example of section 2.2.3. (The tree is explored from the root to leaves and from left to right).

23 consistency checks, the remaining computational effort being used to prove that there is no better solution. See Figure 2.10.

This kind of algorithm has clearly a worst-case behaviour no worse than classical backtracking: both algorithms, in the worst case, will end up trying all possible combinations of values, and testing all the constraints among them. It may actually save effort, as stressed in [26]. Using a pure backtrack search, 44 nodes and 155 consistency checks were needed, the best solution being reached after nine node extensions and 28 consistency checks.

The improvement of the search depends on the bounds b_{inf} and b_{sup}: the higher b_{inf} the more efficient the pruning of useless branches and the lower b_{sup}, the sooner the search will stop. For instance, there are:

- 6 nodes and 24 consistency checks for $b_{inf} = \alpha_3$ and $b_{sup} = 1$;
- 7 nodes and 23 consistency checks for $b_{inf} = 0$ and $b_{sup} = \alpha_3$;
- 3 nodes and 10 consistency checks for $b_{inf} = b_{sup} = \alpha_3$.

It is possible to develop a large class of tree search algorithms (*e.g.* beam search as in [23]) based on the same principles and integrating different enhancements or variants (see [42]). Heuristics for choosing the instatiation ordering of the variables like those proposed by Dechter and Meiri [7] may be used, since they only consider structural characteristics. Dynamic search rearrangement may also be applied: when extending the current instantiation, the variable having the least number of values whose degree of satisfaction is greater than b_{inf} should be chosen first. A variant for assessing the priority of a variable may be to consider the set of the values which are consistent with the current instantiation with a degree greater than $b_{\inf}$:

$$\text{Priority}(x_j/d_1, \ldots, d_k) = \text{Cardinality}\{d_j, \text{Cons}(d_1, \ldots, d_k, d_j) > b_{\inf}\}).$$

For the selection of the value of a variable, the value(s) having the highest degree of satisfaction may be chosen first.

2.3.4 Non-monotonicity in FCSPs

Using the classical CSP approach, the set of solutions shrinks when new constraints are added, and eventually becomes empty in the case of conflicting constraints. In the FCSP framework, adding a new constraint to a problem P may rule out all the previously best solutions if they satisfy the new constraint to a degree lower than $\text{Cons}(P)$. But as long as the new problem (say P') is not totally inconsistent, a new set of best solutions appears that satisfies the new problem to a degree $\text{Cons}(P') \leq \text{Cons}(P)$. Indeed, it holds that:

$$R_1 \otimes \cdots \otimes R_m \otimes R_{m+1} \subseteq R_1 \otimes \cdots \otimes R_m$$

where $\subseteq$ stands for the fuzzy set inclusion, but generally:

$$\{(d_1, \ldots, d_n)/\mu_{R_1 \otimes \cdots \otimes R_m \otimes R_{m+1}}(d_1, \ldots, d_n)\text{Cons}(P')\}$$
$$\not\subseteq \{(d_1, \ldots, d_n)/\mu_{R_1 \otimes \cdots \otimes R_m}(d_1, \ldots, d_n) = \text{Cons}(P)\}.$$

Hence, the set of best solutions does not decrease monotonically when new constraints are added. The non-monotonic behaviour of soft constraints has been noticed by Satoh [41]. The type of non-monotonicity at work here is the same as the one captured by possibilistic logic [16] and appears only in the presence of inconsistency. It has been precisely characterized by Benfehrat *et al.* [1] as the class of preferential inference relations satisfying the rational monotonicity property of Lehmann [34]. In fact, adding a new constraint may lead to four situations:

- the new constraint is redundant: $R_1 \otimes \cdots \otimes R_m \subseteq R_{m+1}$; the set of best solutions remains unchanged;

- the new constraint is totally compatible with P : $\text{Cons}(P) = \text{Cons}(P')$; the set of best solutions is included in the previous one but may remain unchanged;

- the new constraint is partially inconsistent with P : $\text{Cons}(P') < \text{Cons}(P)$; constraints are implicitly relaxed according to their flexibility and the set of best solutions is not necessarily included in the previous one;

- the new constraint is totally incompatible with P : $\text{Cons}(P') = 0$; the set of best solutions is empty.

In the example of section 2.2.3, the consistency of the problem is α_3 and the set of best solutions consists of a single one, namely $\{(x = 3, y = 3, z = 1)\}$:

- adding the redundant hard constraint $z \leq 3$ does not change the consistency of the problem nor the set of best solutions;

- adding the compatible hard constraint $y + z = 4$ neither changes the consistency of the problem nor the set of best solutions; however, the satisfaction degrees of other instantiations decreases (*e.g.* the satisfaction degrees of $(x = 4, y = 1, z = 2)$ become 0 instead of α_2);

- adding the hard constraint $y + z = 3$, the consistency of the problem becomes α_2 and the new set of best solutions is $\{(x = 4, y = 1, z = 2), (x = 4, y = 2, z = 1), (x = 4, y = 0, z = 3)\}$;

- adding the hard constraint $x + y = 3$, the problem becomes totally inconsistent.

As a consequence of this non-monotonic behaviour, the problem of solution maintenance in FCSPs appears to be more complex than in classical CSPs: pruned branches in a previous search through the tree have to be developed contrary to the method proposed by Van Hentenryck [44]. The question of relaxing or deleting a constraint is not separately considered in the FCSP model, since the relaxation capacities of the constraints are supposed to be explicitly represented by means of preference among values and priority degrees. In other terms, in the FCSP model, the allowed weakening and deletion of constraints are already captured by the flexibility of the constraints (as far as preferences remain unchanged). On the contrary, when constraints have to dynamically be added or strengthened, *e.g.* the priority of a constraint (resp. the satisfaction degree of a value) increase (resp. decrease), the non-monotonicity phenomena described above takes place.

2.4   Local Consistency in FCSPs

In the classical CSP framework, local consistency techniques can be used to improve the efficiency of the search algorithms. The most common techniques enforce arc-consistency (*e.g.* the AC-3 algorithm proposed by Mackworth [35] which is a generalization and a simplification of the earlier 'filtering' algorithm by Waltz [45]) or path-consistency (*e.g.* the PC-2 algorithm proposed also by Mackworth [35]). These preliminary notions of local consistency have been generalized by the concept of k-consistency [24]. Actually, the theoretical foundations of the FCSP framework, as presented in the previous sections, allow us to easily extend all these definitions and the associated algorithms (*e.g.* AC3) [21].

2.4.1 Local Consistency for Fuzzy Constraints

- **k-consistency**

  A classical CSP is said to be k-consistent if any consistent instantiation of $k-1$ variables can be extended to a consistent instantiation involving any k^{th} variable. A FCSP is said to be k-consistent if any instantiation of $k-1$ variables can be extended to a partially consistent instantiation involving any k^{th} variable, and this instantiation must be as consistent as the instantiation of $k-1$ variables. Formally, a FCSP is k-consistent if and only if:

  $$\forall\{x_{j_1}, \ldots, x_{j_{k_1}}\} \subseteq X, \forall x_{j_k} \in X \text{ such as } x_{j_k} \notin \{x_{j_1}, \ldots, x_{j_{k-1}}\},$$

  $$\forall(d_{j_1}, \ldots, d_{j_{k-1}}) \in D_{j_1} \times \cdots \times D_{j_{k-1}},$$

  $$\exists d_{j_k} \in D_{j_k} \text{ such as } \mathrm{Cons}(d_{j_1}, \ldots, d_{j_{k-1}}, d_{j_k}) = \mathrm{Cons}(d_{j_1}, \ldots, d_{j_{k-1}}).$$

  A FCSP is strongly k-consistent if it is j-consistent for every $j \le k$. According to this definition, a necessary and sufficient condition for k-consistency is:

  $$\forall\{x_{j_1}, \ldots, x_{j_{k-1}}\} \subseteq X, \forall x_{j_k} \in X \text{ such as } x_{j_k} \notin \{x_{j_1}, \ldots, x_{j_{k-1}}\},$$

  $$\otimes_{\{R_i, V(R_i) \subseteq \{x_{j_1}, \ldots, x_{j_{k-1}}\}\}} R_i = [\otimes_{\{R_i, V(R_i) \subseteq \{x_{j_1}, \ldots, x_{j_k}\}\}} R_i]^{\downarrow\{x_{j_1}, \ldots, x_{j_{k-1}}\}}$$

- **Arc-consistency**

  A network of fuzzy constraints is arc-consistent (or equivalently 2-consistent) if and only if:

  $$\forall x_i \in X, \forall x_j \in X \text{ such as } x_i \ne x_j, R_i \subseteq [R_{ij} \otimes R_j]^{\downarrow\{x_i\}},$$

  R_i, R_j and R_{ij} respectively denoting the relation associated to the unary constraint on x_i, x_j and the binary constraint relating x_i and x_j. This condition is a generalization to fuzzy relations of the usual condition of arc consistency.

- **3-consistency**

  A network of fuzzy constraints is 3-consistent if and only if:

  $$\forall\{x_i, x_j\} \subseteq X, \forall x_k \in X-\{x_i, x_j\}, R_i \otimes R_{ij} \otimes R_j \subseteq [R_{ik} \otimes R_k \otimes R_{kj}]^{\downarrow\{x_i, x_j\}}.$$

  Combination and projection of fuzzy relations share all properties of combination and projection of crisp relations. Hence basic results in classical CSPs extend right away to the FCSP framework, like Mackworth's theorem relating 3-consistency to path consistency [35] and the backtrack-free sufficient conditions of Freuder [25] or Dechter [6]. See Fargier [21] for detailed proofs. In fact, properties which cannot be used in FCSPs are the monotonicity, as outlined previously, and properties related to the negation: if R_i represents a flexible constraint C_i and R_i^c the constraint $not(C_i)$, $R_i \otimes R_i^c = \emptyset$ does not hold generally. As soon as the description language does not include the negation of constraints, all results in classical static CSPs still hold in FCSPs.

2.4.2   Network Consistency Algorithms for FCSPs

All the classical filtering algorithms extend to FCSPs as well. The following
algorithm is an extension of AC3 which ensures arc-consistency in a network of
fuzzy constraints $(L = [0, 1])$. It provides in addition an upper approximation
of the overall consistency degree $\text{Cons}(P)$.

Procedure FAC-3
Cons-P-sup $\leftarrow 1$
$Q \leftarrow \{(i, j)/\exists R_h \in R \text{ s.t. } V(R_h) = \{x_i, x_j\}, i \neq j\}$
While Q not empty, do
select and delete any arc (k, m) from Q
if Revise$(k, m, \text{Cons} - P - \text{sup})$ do
$\qquad Q \leftarrow Q \cup \{(i, k)/\exists R_h \in R \text{ s.t. } V(R_h) = \{x_i, x_k\}, i \neq k, i \neq m\}$
return Cons-P-sup.

Procedure Revise $(i, j, \text{Cons-P-sup})$
Changed $\leftarrow$ false
Height $\leftarrow 0$
for each d_i in D_i do
$\qquad$ for each d_j in D_j do
$\qquad\qquad$ new-degree $\leftarrow \min(\mu_{R_i}(d_i), \mu_{R_{ij}}(d_i, d_j), \mu_{R_j}(d_j))$
$\qquad\qquad$ Height $\leftarrow \max(\text{new-degree, Height})$
$\qquad\qquad$ if new-degree $= 0$, delete d_i from D_i.
$\qquad\qquad$ if new-degree $\neq \mu_{R_i}(d_i)$, do
$\qquad\qquad\qquad$ Changed $\leftarrow$ true
$\qquad\qquad\qquad \mu_{R_i}(d_i) \leftarrow$ new-degree.
Cons-P-sup $\leftarrow \min (\text{Cons-P-sup, Height})$
return Changed.

Other classical filtering algorithms can be straightforwardly extended by
changing the 'revise' procedure. For instance, PC2 is to be performed as
defined by Mackworth [35], replacing the updating pattern by its fuzzy coun-
terpart:

$$R_{ij} \leftarrow R_{ij} \otimes [R_{ik} \otimes R_k \otimes R_{kj}]^{\downarrow \{x_i, x_j\}}$$

$\otimes$ and $\downarrow$ denoting the fuzzy conjunctive combination and projection.

When performed on a FCSP, the complexity of any classical filtering algorithm must at most be augmented by a factor p, if p is the number of
different levels of satisfaction used to describe the flexibility of the problem.
This result can be intuitively understood since a FCSP is equivalent to p
classical CSPs, as outlined in section 2.3. In fact, consider FAC3, d denoting
the maximal cardinality of domains, p being the number of different levels
of satisfaction, m the number of binary constraints and d the maximal cardinality of domains. The cost of the 'revise' procedure remains unchanged.
The list Q is increased when a call to revise has succeeded. In classical CSPs,
revise(x_i, x_j) is called with success at most d times. In a FCSP, revise(x_i, x_j)

is called with success at most $p * d$ times since each possible value for x_i may have its degree diminished at most p times: combination is idempotent, and leads to decreased satisfaction levels without generating levels other than the original ones. Successful calls of 'revise' concern at least one of the d possible values for x_i. Hence, the theoretical complexity of FAC3 is $O(pd^3m)$. More sophisticated algorithms, like AC4 [37], can also be easily adapted to FCSP (see Fargier [20]). In [43, 17, 32] the use of hypergraph structures appears to be a generalization of tree clustering [8] to fuzzy constraints.

2.5 Conclusion

The rich expressive power of possibility theory provides a general and unified framework for the representation and management of flexible constraints involving preferences on values as well as prioritized constraints. The FCSP formalism, which is a generalization of classical CSPs, nevertheless offers a large variety of efficient problem solving tools: most classical CSP algorithms easily extend, as well as most of the CSP theoretical results and their applications (*e.g.* tree clustering). This is due to the fact that FCSPs are not additive, but solely based on commensurate orderings, so that all useful properties of the Boolean structure underlying classical CSPs remain valid. The FCSP framework is currently applied to constraint-based approaches in job-shop scheduling [11] where flexible constraints and uncertain parameters are usual features.

As it turns out, explicitly taking the flexibility of the problem into account does not drastically increase the worst-case computational cost of the search procedure; the complexity of filtering procedures may be increased by a factor reflecting the number of different levels used to describe flexibility in the application under concern. Moreover, the problem of finding a feasible solution is changed into an optimization problem of the bottleneck kind, to which Branch and Bound procedures may apply. Of course, in practice, finding an optimal solution is generally more computationally expensive than finding a feasible solution. But experiments carried out in the area of scheduling indicate that the first feasible solution found in the FCSP framework is often obtained more quickly than when preferences are neglected [21]. Moreover, the FCSP approach bypasses empirical relaxation techniques which are needed when a set of constraint is globally unfeasible. Constraint relaxation often happens to be more expensive, difficult to formulate, and suboptimal. On the contrary, the FCSP approach can handle partially inconsistent problems. A solution (the instantiation with the maximal satisfaction degree) will be provided as long as the problem is not totally inconsistent. Hence fuzzy constraints are also useful to guide the search procedure towards 'interesting' solutions. Theoretical extensions of the framework are planned with a view to developing computational tools for handling refinements of the global minimum-based satisfaction ordering used here that may be judged as not enough discriminant [21, 12]. Finally, this formalism suggests a non-monotonic framework for dynamic CSPs, when for instance in computer aided-design, default constraints which are used in a first step analysis are then dynamically modified by the designer.

Moreover, the framework offered by possibility theory enables us to represent ill-known parameters, whose precise value is neither accessible nor under our control, under the form of so-called possibility distributions (where the possible values are rank-ordered according to their level of plausibility). Ill-known parameters contrast with decision variables on which a decision-maker has control. It is shown in [13] that constraints whose satisfaction depends on these ill-known parameters can be represented in the setting of possibility theory as well. In the presence of ill-known parameters, robust solutions should be searched for, such that the constraints are satisfied whatever the values of these ill-known parameters. Possibility theory implements this idea in a flexible way.

References

[1] Benfehrat S., Dubois D. and Prade H. (1992) Representing default rules in possibilistic logic. *Proc. of the 3rd Inter. Conf. on Principles of Knowledge Representation and Reasoning (KR'92)* Nebel B, Rich C., Swartout W. (eds) Cambridge, MA, Morgan & Kaufmann, San Mateo, CA, 673–684.

[2] Borning A., Maher M., Marindale A. and Wilson M. (1989) Constraint hierarchies and logic programming. *Proc. of the Inter. Conf. on Logic Programming* Lisbon, Portugal, June 149–164.

[3] Bowen J., Lai R. and Bahler D. (1992) Fuzzy semantics and fuzzy constraint networks. *Proc. of the 1st IEEE Conf. on Fuzzy Systems (FUZZ-IEEE'92)* San Fransisco, 1009–1016.

[4] Bowen J., Lai R. and Bahler D. (1992) Lexical imprecision in fuzzy constriant networks. *Proc. of the National Conf. on Artificial Intelligence (AAAI'92)* 616–620.

[5] Brewka G., Guesgen H. and Hertzberg J. (1992) Constraint relaxation and nonmonotonic reasoning. German National Research Center (GMD), Report TR-92-02.

[6] Dechter R. (1992) From local to global consistency. *Artificial Intelligence* 55: 87–107.

[7] Dechter R. and Meiri I. (1989) Experimental evaluation of preprocessing techniques in Constraint Satisfaction Problems. *Proc. of the Inter. Joint Conf. on Artificial Intelligence (IJCAI'89)* Detroit, MI, 271–277.

[8] Dechter R. and Pearl J. (1989) Tree clustering for constraint networks. *Artificial Intelligence* 38: 353–366.

[9] Descottes Y. and Latombe J.C. (1985) Making compromises among antagonist constraints. *Artificial Intelligence* 27: 149–164.

[10] Dubois D., Fargier H. and Prade H. (1990) The calculus of fuzzy restriction as a basis for flexible constraint satisfaction. *Proc. of the 2nd IEEE Inter. Conf. on Fuzzy Systems (FUZZ-IEEE'93)* San Fransico, CA, March 28-April 1st, 1131–1136.

[11] Dubois D., Fargier H. and Prade H. (1993) The use fuzzy constraints in job-shop scheduling. *Proc. of the IJCAI'93* Workshop on Knowledge-Based Production Planning Scheduling and Control, Chambry, France, Aug. 29, 101–112. Extended version to appear in the J. of Intelligent Manufacturing.

[12] Dubois D., Fargier H. and Prade H. (1995) Refinements to the maximin approach to decision-making in fuzzy environment. *Fuzzy Sets and Systems*, to appear.

[13] Dubois D., Fargier H. and Prade H. (1995) Possibility theory in constraint satisfaction problems: Handling priority, preference and uncertainty. *Applied Intelligence*, to appear.

[14] Dubois D., Prade H. (with the collaboration of Farreny H., Martin-Clouaire R., Testemale C.) (1988) *Possibility Theory An Approach to Computerized Processing of Uncertainty.* Plenum Press, New York.

[15] Dubois D. and Prade H. (1989) Processing fuzzy temporal knowledge. *IEEE Trans. on Systems, Man and Cybernetics* (19) 4: 729–744.

[16] Dubois D. and Prade H. (1991) Possibilistic logic, preferential models, non-monotonicity and related issues. *Proc. of the Inter. Joint Conf. on Artificial Intelligence (IJCAI'91)*, Sydney, Australia, Aug. 24–30, 419–424.

[17] Dubois D. and Prade H. (1991) Inference in possibilistic hypergraphs. In: *Uncertainty in Knowledge Bases*, (Proc. of the 3rd Inter. Conf. on Information Processing and Management of Uncertainty in Knowledge-Based Systems (IPMU'90), Paris, France, July 1990) (Bouchon-Meunier B, Yager R.R., Zadeh L.A. (eds), *Lecture Notes in Computer Science* (521) Springer Verlag, Berlin, 250–259.

[18] Dubois D. and Prade H. (1993) Tolerant fuzzy pattern matching: An introduction. *J. of Fuzzy Logic and Intelligent Systems* (3) 2: 3–17 Seoul, Korea.

[19] Faltings B., Haroud D. and Smith I. (1992) Dynamic constraint propagation with continuous variables. *Proc. of the Europ. Conf. on Artificial Intelligence (ECAI'92)* 754–758.

[20] Fargier H. (1992) Problèmes de satisfaction de contraintes floues. Technical Report IRIT/92-29-R, I.R.I.T., Universit P. Sabatier, Toulouse, France.

[21] Fargier H. (1994) *Problèmes de satisfaction de contraintes flexibles Application l'ordonnancement de production.* Thèse de l'Université P. Sabatier, Toulouse, France.

[22] Fargier H., Lang J. and Schiex T. (1993) Selecting preferred solutions in fuzzy constraint satisfaction problems. *Proc. of the 1st Europ. Conf. on Fuzzy Information Technologies (EUFIT'93)* Aachen, Germany, Sept. 7-10, Published by ELITE-Foundation, Aachen, 1128–1134.

[23] Fox M., Allen B. and Strohm (1982) Job-shop scheduling: An investigation in constraint-directed reasoning. *Proc. of the National Conf on Artificial Intelligence (AAAI'82)* Pittsburgh, USA, 155–158.

[24] Freuder E.C. (1978) Synthetising constraint expressions. *Communications of the ACM* (21) 11: 958–966.

[25] Freuder E.C. (1982) A sufficient condition for backtrack-free search. *J. of the ACM* (32) 4: 755–761.

[26] Freuder E.C. (1989) Partial constraint satisfaction. *Proc. of the Inter. Joint Conf. on Artificial Intelligence (IJCAI'89)* Detroit, MI, 278–283.

[27] Freuder E.C. and Snow P. (1990) Improved relaxation and search methods for approximate constraint satisfaction with a maximin criterion. *Proc. of the 8th Biennial Conf. of the Canadian Society for Computational Studies of Intelligence* Ontario, Canada, May 22–25, 227–230.

[28] Freuder E.C. and Wallace R. (1992) Partial constraint satisfaction. *Artificial Intelligence* 58: 21–71.

[29] Goguen J.A. (1967) L-fuzzy sets. *J. of Mathematical Analysis and Application* 18: 145–174.

[30] Guan Q. and Friedrich G. (1992) Extending constraint satisfaction problem solving in structural design. *Proc. of the 5th Inter. Conf. IEA/AIE* Paderborn, Germany, June, 341–350.

[31] Hummel R. and Zucker S. (1983) On the foundations of relaxation labelling processes. *IEEE Trans. on Pattern Analysis and Machine Intelligence* (5) 3: 267–287.

[32] Kruse R. and Schwecke E. (1990) Fuzzy reasoning in a multidimensional space of hypotheses. *Int. J. of Approximate Reasoning* 4: 47–68.

[33] Lang J. (1991) Possibilistic logic as a logical framework for min-max discrete optimization problems and prioritized constraints. *Proc. of the Inter. Workshop on Fundamentals of Artificial Intelligence Research (FAIR'91)* Smolenice, Czechoslovakia, Sept. 8-12, Lecture Notes in *Computer Science* (535) Springer Verlag, Berlin, 112–126.

[34] Lehmann D. (1989) What does a conditional knowledge base entail?. *Proc. of the 1st Inter. Conf. on Principles of Knowledge Representation and Reasoning (KR'89)* Toronto, 212–221.

[35] Mackworth A.K. (1977) Consistency in networks of relations. *Artificial Intelligence* 8: 99–118.

[36] Martin-Clouaire R. (1992) Dealing with soft constraints in a constraint satisfaction problem. *Proc. of the Inter. Conf. on Information Processing and Management of Uncertainty in Knowledge-Based Systems (IPMU'92)* Mallorca, Spain, July 6–10, 37–40.

[37] Mohr R. and Henderson T. (1986) Arc and path consistency revisited. *Artificial Intelligence* 28: 225–233.

[38] Moulin H. (1988) *Axioms of Cooperative Decision Making.* Cambridge University Press, Cambridge, MA.

[39] Rosenfeld A., Hummel R.A. and Zucker S.W. (1976) Scene labelling by relaxation operations. *IEEE Trans. on Systems, Man and Cybernetics* 6: 420–433.

[40] Sadeh N. (1991) Look-ahead techniques for micro-opportunistic job shop scheduling. Carnegie Mellon University, Report CS-91-102.

[41] Satoh K. (1990) Formalizing soft constraint by interpretation ordering. *Proc. of the Europ. Conf. on Artificial Intelligence (ECAI'90)* Stockhom, Sweden, 585–590.

[42] Schiex T. (1992) Possibilistic constraint satisfaction problems or how to handle soft constraints. *Proc. of the 8th Conf. on Uncertainty in Artificial Intelligence* Dubois D., Wellman M.P., D'Ambrosio B. and Smets P. (eds) Stanford, CA, July 17-19, 1992, Morgan & Kaufmann, San Mateo, CA, 268–275.

[43] Shafer G. and Shenoy P. (1990) Axioms for probability and belief function preparation. In: *Uncertainty in Artificial Intelligence* (4) Shachter R.D., Levitt T.S., Kanal L.N. and Lemmer S.F. (eds) North-Holland, Amsterdam, 169–198.

[44] Van Hentenryck P. (1990) Incremental constraint satisfaction in logic programming. *Proc. ICLP 90* 189–202.

[45] Waltz D. (1975) Understanding line drawings of scenes with shadows. In: *The Psychology of Computer Vision* Winston P.H. (ed) McGraw-Hill, New York, 19–92.

[46] Yager R.R. (1989) Some extensions of constraint propagation of label sets. *Int. J. of Approximate Reasoning* 3: 417–435.

[47] Zadeh L.A. (1975) Calculus of fuzzy restrictions. In: *Fuzzy Sets and Their Applications to Cognitive and Decision Processes* Zadeh L.A. *et al.* (eds) Academic Press, New-York, 1–39.

[48] Zadeh L.A. (1978) Fuzzy sets as a basis for a theory of possibility. *Fuzzy Sets and Systems* 1: 3–28.

3

Knowledge from Data using Fril and Fuzzy Methods

J.F. Baldwin

3.1 Introduction

Human knowledge is obtained by induction — we generalize from known cases to new cases which are similar. We use induction to make decisions and to classify, to draw conclusions, to form useful rules of thumb, to make models and to hypothesize. There is no deductive logic which we can use to validate inductive arguments.

We are still at an early stage of discovering methods which we can use to make inductions. In this chapter we propose an approach using fuzzy methods. It is a new approach and does not use the existing methods of fuzzy logic or fuzzy control. In order to persuade the reader that the basic ideas are worthy of consideration, we demonstrate the effectiveness of such an approach for some varying types of examples and compare the performance with some more classical AI techniques, and also the use of artificial neural nets.

For any particular problem there is no correct solution. The next number in the sequence 1, 2, 3, 4 could be any number, but if we believe that these numbers were not simply chosen at random and that there is a pattern which we can use to guess the next number, then we will choose 5. Similarly if we are shown a group of faces labelled male and a group labelled female we can suppose that these classifications have not been randomly selected, and that the classification will effect various characteristics of the faces. We can use these varying characteristics of the faces to help choose a classification for a new unclassified face. What characteristics should we use and what weights of importance should we give to them? A characteristic may be a simple feature or a combination of features. Some characteristics will be more discriminatory than others. Those with greater powers of discrimination would be thought to be more important. Some characteristics may more strongly suggest one classification over others, while another characteristic may suggest something different. How do we combine such suggestions or evidences or supports or beliefs? It is not enough to simply find a rule which will predict the classifications of the known cases and then use this rule for the unknown cases. There will not be a unique rule with this property, so how do we choose one rule from a number of rules? In the case of neural nets we could choose different architectures which would give similar success on the learning set of known cases. Each of these architectures could give a different solution for the unknown cases. Which architecture should we use? Why should we have more confidence in the generalization given by one architecture as compared to another? The problem may not be a classification problem. It could be the prediction of a point value as in the case of function approximation. Once again, an inductive procedure which gives very close agreement with the known cases will not be unique.

Science tends to prefer a simple to a more complex theory and likewise we will prefer simple to complex rules. But what do we mean by simple? We must test what ever inductive procedure we choose to classify or predict. This testing can be Poperian in style — would try to select what appears to be a difficult case, determine the classification or prediction for this case from experiment and see how well the inductive inference performs. A rule for classification which says that the classification is C if the example is one in the example set with classification C has no powers of generalization. It can say nothing about unknown cases.

Why should we be so interested to discover ways for computers to make inductions? Data is everywhere — we can easily collect large databases and store them for future use. Large volume data from various fields such as medicine, science and engineering, finance and economics, marketing, social welfare, education, advertising, law, the arts, etc. is there for the collecting. We store it away in large databases in the hope that it will be useful. To be useful we must be able to discover relationships between the various attributes of the database, find rules to predict values of one attribute when the others are known for cases which are not explicitly present in the database. Databases will not have all its entries correct, there will be errors. Some entries will not be known exactly. We will not be sure of the exact context for which the database is relevant.

How can we extract knowledge from these databases where we will understand knowledge to be useful relationships and useful rules? We may wish to report this knowledge in the form of a natural language summary. Of course, at present we have methods for doing something like this. Statistical methods and other numerical techniques can extract some useful information. We are not suggesting that we should throw these methods away. But we are asking for something different and something more. There should be a conversation between the human user and the computer. The computer can discover rules, relationships, perform classifications, make decisions, predict attribute values but to do this there will be assumptions, a particular point of view will be used and there will be a certain confidence in the results. The user must determine his/her willingness to accept such decisions. This may involve asking for additional predictions, testing the rules against another set of data, asking what if questions, determining the sensitivity of the predictions against possible changes in the data, etc.

The computer is acting as an accountant with discovery powers and the human user is acting as an intelligent agent with limited ability to handle large data sets effectively. The user might suggest relationships or rules which the computer can check. The user may wish to know the truth of certain propositions for which the answer can quite easily be determined using simple computational procedures. The computer should be able to decide which procedures to use, in what order and how to combine the results to give an answer. The answer may be a probability of the proposition being true, a truth value or a more expressive statement. Suppose the user wants to know the truth of the proposition 'chemotherapy can help bone cancer patients'. The computer would use various databases through out the world to provide an answer to this question. The answer would not be a simple true or false, the answer would be inappropriate if it were simply a probability. Perhaps the computer could discover some interesting relations between the success of the treatment and certain characteristics of the patient. For example, in many cases in which the bone cancer is reasonably contained at the beginning of the treatment and otherwise the patient is young and in good health, there is a good chance of recovery for many years. To obtain this answer we require much more than a simple statistical numerical technique or a simple neural net. Yet for small databases human intelligence could extract such a statement. The combined human/computer performing as partners, improvising according to each other's skills, could make discoveries more important than those able to be obtained by either working alone.

The conclusion of this discussion is that we require a data browser in the form of a human/computer partnership where the human is the actual interested party and the computer has the appropriate software for making inductions and performing AI tasks. The net provides access to databases throughout the world. The net also provides access to other computers with other methods and also to other experts and interested parties. Teams of humans and computers with different software could combine, join forces in this discovery process. This combination would evolve as time goes by in trying to discover something in particular for a specific user. The data browser should be able to handle this network of human and computer activity.

The data browser is written in the AI programming language FRIL, [15, 12].

3.2   An Intelligent Data Browser in Fril

Data is provided in the form of a database. The database entries are attribute values taken from the domain of the attribute. These values can be specific, a sequence of possible specific values including a range of values if the domain is continuous or a fuzzy set on the domain. A query can ask for any attribute value given information concerning the other attribute values. The solution may not be directly found in the database. It is then inferred by using the database as a set of examples to derive Fril rules.

We discuss an example to illustrate this approach. Consider the following database:

FUNCTION	Y	X1	X2	X3	X4	X5
F	2.09	0.123	1.245	[1,2]	**small**	3.68
	etc.					

It represents a function F whose mapping is known at a finite set of points. The range $[1, 2]$ indicates a value of $X3$ lying in this range and small is a fuzzy set defined on the domain of $X4$. The query might be to find the value of F, *i.e.* Y, when $X1= 1$, $X2 \in [0, 0.2]$, $X3$ is **almost zero**, $X4 = 0.7$, $X5 = $ **average** where **almost zero** and **average** are fuzzy sets on the appropriate domains. We will further suppose that there is no row in the database corresponding exactly to these Xi attribute values. To answer the query we must generalize from those mapping instances given to provide the particular mapping asked for.

This required generalization could be achieved using a multi-dimensional linear interpolation with respect to the known attribute values of the nearest instances. Alternatively, we can use all the data, or even a subset of the data representing near cases, to form rules which we can then use to infer the required solution. We choose this latter approach which, of course, will be an effective multi-dimensional interpolation.

What form of rules should we use and how do we derive such rules? In Fril we can use various forms of rules:

1. Basic Fril Rule;

2. Extended Fril Rule;

3. Evidential Logic Rule;

4. Causal Rule.

A full treatment of Fril is given in the book [12]. The basic Fril rule is of the form:

$$(\text{head IF body}) \quad : \quad ((x1\ x2)(y1\ y2))$$

which says $\Pr(\text{head} \mid \text{body}) \in [x1, x2]$ and $\Pr(\text{head} \mid \neg\ \text{body}) \in [y1, y2]$. The head of the rule is a list whose first element is a predicate, for example (weight of X is **heavy**) and the body is a conjunction of similar terms, for example (height of X is **tall**)(shoulders of X are **broad**). If the support pairs, $((x1x2)(y1y2))$, are missing then the rule is an ordinary logic rule, which reads as 'someone is heavy' IF **tall** and has **broad** shoulders. This is an implication rule. If the support pairs are $((1\ 1)(0\ 0))$ then it is an equivalence rule. The extended rule allows more support pairs so that the body of the rule can be given a set of mutually exclusive instantiations.

The evidential logic rule takes the form:

> (head
>> (evlog **fuzzy_set** (conjunction of weighted terms)))
>> :$((x1x2)(y1y2))$.

The head has the same form as before and each weighted term in the body takes the form

> $(\ldots)w$

where w is the weight. The fuzzy set **fuzzy_set** is a filter which modifies the degree of satisfaction of the body.

An example of this rule is:

> ((suitability_as_sports_administrator of person X is **good**)
>> (evlog **most** (
>>> (organizational_ability of X is **high**) 0.1
>>> (fitness of X is **fairly_good**) 0.1
>>> (qualifications X are **applicable**) 0.2
>>> (concentration X is **long**) 0.3
>>> (numeracy of X is **very_good**) 0.2
>>> (availability of X is **flexible**) 0.1))) : $((1\ 1)(0\ 0))$.

For the above database example we will use a collection of rules of the

> Y is $\mathbf{f_i}$ IFF X1 is $\mathbf{g_{i1}} \wedge X2$ is $\mathbf{g_{i2}} \wedge \ldots$ X5 is $\mathbf{g_{i5}}$
> for i = 1, ..., n
> where $\mathbf{f_i}$ is a fuzzy set on the domain of Y and $\mathbf{g_{ij}}$ are fuzzy sets on
> the domain of Xij.

The fuzzy sets $\mathbf{f_i}$ are chosen to be a mutually exclusive set of fuzzy sets covering the domain of Y where we define mutually exclusive to mean $\sum_i \mu_{f_i}(y) = 1$ for any $y \in Y$. The fuzzy sets $\mathbf{g_{ij}}$ are then chosen using a method based on mass assignment theory, as explained later. This method of choosing the $\mathbf{g_{ij}}$ assumes that we can work on each attribute separately. By taking the relevant statistics of the occurrence of each attribute value into account we compensate for not considering the cross product attribute space. This compensation may not be sufficient to provide the required accuracy, and we may have to introduce additional attributes to provide further compensation. The accuracy of the rules is tested on the original database. For this purpose we can split the database into a learning database subset and a testing subset. The rules are learned using the learning subset and then tested using the testing subset.

Additional attributes are chosen as algebraic combinations of the present attributes using genetic programming or some other method. This will not always be required but is useful in the more difficult cases. Additional attributes are redundant if we used the complete multi-dimensional space to derive our fuzzy sets.

The rules are thought of as a collection of rules and the mass assignment theory provides an inference mechanism which allows the value of Y to be inferred from this collection and the input data for the Xi values. The inference rule is of the form:

Y is $\mathbf{f}$ where $\mathbf{f} = \sum_i \theta_i f_i$.

The $\{\theta i\}$ in this inference rule are derived using point semantic unification, which is explained later, and represents the degree to which the body of each rule is satisfied by the input data.

This prediction is in the form of a fuzzy set resulting from taking the weighted sum of the fuzzy sets associated with the heads of the rules. We now take the expected value of this fuzzy set as our final prediction, *i.e.*

Y is $E(\mathbf{f})$.

This expected value is determined using mass assignment theory as explained later. It is, in fact, the expected value of the least prejudiced probability distribution associated with the fuzzy set $\mathbf{f}$.

We should also note that the rules in the collection are equivalence rules and not implications. Furthermore, the inference rule is not applied to each rule separately, as would be the case with an ordinary logic inference, but the rules are taken together.

For functions of one variable, this approach to predicting values of the function is equivalent to linear interpolation if a set of mutually exclusive triangular fuzzy sets are chosen on the Y domain.

The data browser can handle classification problems in a similar manner. Consider the database:

CLASSIFICATION	Class	X	Y
Colour	colour_1	**small**	**about_0**
	colour_2	6	[6. 10]
	etc.		

The class attribute can take values from {legal, not legal} and the X and Y attributes can take values from the domains [-10, 10], [-12, 12], respectively. The attribute values **small** and **about_0** are fuzzy sets defined on these domains and also a range of values can be used.

If we have data inputs for X and Y not corresponding to any row in the table then we form rules to aid with the classification. This time we might use rules of the form:

> Y is colour_i IFF X is $g_{i1} \wedge Y$ is g_{i2}
> for $i = 1, \ldots, m$
> where g_{ij} is a fuzzy set on the domain of X and h_{ij} are fuzzy
> sets on the domain of Y.

The inference rule would then provide a solution of the form:

> Y is colour_1 : θi
> for $i = 1, \ldots, m$.

The final classification is the one with the higher θ value, although if the difference is too small then one could choose to make no decision. A more general decision theoretic approach using utility theory could also be used. The fuzzy sets in the rules are determined as before, and the consideration of using additional attributes as algebraic combinations of the present ones also applies.

Alternatively, we could use an evidential logic rule instead of the above equivalence rules. This allows the features in the body to have importances and only most features need be satisfied. The features are of the form (feature value is in f), where f is a fuzzy set defined on the domain of the feature. This feature will be automatically extracted from the database. The weight associated with the feature will also be determined automatically using a form of discrimination analysis, as explained in a later section. The feature can be a simple attribute of the database or a combination of such attributes. The form of the combination will depend on the type of domain for each of the attributes in the combination. This will be explained further in later sections.

In the data browser the user can choose what sort of rules to use and test the accuracy of the derived rules. The user can also suggest changes in the rules, propose new attributes, change the derived fuzzy sets. This can be done easily through a simple menu driven interface. This computer/user interaction should help the user to have a much better comprehension of the data and the combination will provide more intelligence than if either the user or the computer were acting alone.

Another use of the data browser is to handle the sort of queries which are normally associated with Zadeh's PRUF language. These queries require combinations of defined procedures to provide the answer, [10]. Generalization of the database may not be necessary but the correct procedures and order of operation of the procedures for efficient computation must be chosen automatically from the knowledge of the query. The query can be asked in the style of natural language. The query is analysed and a problem generator used to provide the Fril query necessary to provide the answer.

Suppose we have databases with information about students in a university. We might wish to ask any of the following queries:

What proportion of first year students came with **good** school leaving qualifications?

Is it true that **most** students who graduate with **high** honours performed **well** over all years?

It is **generally thought** that the **hard** subject areas are **less popular** with the students. Is this true?

To answer these questions certain rules are required to define certain terms. For example, what do we mean by good school leaving qualification. The data browser may have a definition, but if not would converse with the user to acquire one. If the data browser does have such a definition the user may disagree with it so some form of conversation is still required.

The last query introduces the controversial concept of what is a hard subject. Again the solution is determined by considering topics within a given subject as well as more general subjects. Physics is thought be hard, general relativity is harder than special relativity within the Physics field. Some form of conversation with the user is required to determine the exact context in which to answer the query.

Once the query is well understood the choosing of the procedures to answer such a query is relatively simple. The ordering is not quite so easy. The ordering should be such that the size of the relevant databases is quickly reduced. Heuristics can be used for this purpose.

3.3   Essentials of Mass Assignment Theory

In this section we explain using a simple example the essentials of mass assignment theory [1, 2, 3, 4, 5, 6, 7, 11] for understanding the next section.

You are told that a weighted dice is thrown and the value is **small** where **small** is a fuzzy set defined as

small = 1 /1 + 2 / 1 + 3 / 0.4.

The prior for the dice is

$1 : 0.1, 2 : 0.2, 3 : 0.3, 4 : 0.2, 5 : 0.1, 6 : 0.1.$

Can we derive the distribution $\Pr(\text{dice is } i \mid \text{dice is } \textbf{small})$?

What is $\Pr(\text{dice is } \textbf{about\_2} \mid \text{dice is } \textbf{small})$ where **about_2** is a fuzzy set defined as

about_2 = 1 / 0.3 + 2 / 1 + 3 / 0.3.

The mass assignment for **small** in this example is equivalent to the basic probability assignment of the Dempster Shafer theory. It is a random set which we can write as

$m_{rmsmall} = \{1, 2\} : 0.6, \{1, 2, 3\} : 0.4.$

This represents a probability distribution on the power set of S, where S is the set of possible dice values, and a family of probability distributions on the set S. This family arises since we can allocate a mass associated with a set of elements to the individual elements of that set in many ways. The semantics of mass assignments can be understood from the voting model interpretation of a fuzzy set [1, 2, 3, 4, 5, 6, 7]. Related papers are [13, 14, 16, 17, 18].

To obtain the least prejudiced distribution we allocate a mass associated with a set of dice values to the individual members of that set in proportion to the probabilities of the prior. Thus the mass of 0.6 associated with $\{1, 2\}$ is allocated as 0.2 to 1 and 0.4 to 2. The mass of 0.4 associated with $\{1, 2, 3\}$ is allocated as $0.4(1/6) = 0.0667$ to 1, $0.4(2/6) = 0.1333$ to 2 and $0.4(3/6) = 0.2$ to 3. Thus the least prejudiced distribution for dice given dice is **small** is given by

$$\text{lpd}(\text{Dice} \mid \textbf{small}) = 1 : 0.2667, 2 : 0.5333, 3 : 0.2.$$

We take this as representing $\Pr(\text{dice is } i \mid \text{dice is } \textbf{small})$.

The above calculation is equivalent to

$$\Pr(\text{dice is } i \mid \text{dice is } \textbf{small}) = 0.6\Pr(i \mid \{1, 2\}) + 0.4\Pr(i \mid \{1, 2, 3\})$$
where
$$Pr(i|\{\ldots\}) = \frac{Pr(\{\ldots\}|i)Pr(i)}{Pr(\{\ldots\})} \qquad \text{Bayes Theorem.}$$

Thus the updating procedure when given a fuzzy set is an extended Bayes' theorem.

Fril can determine the distribution for any discrete fuzzy set and the least prejudiced density for any continuous fuzzy set.

We will use this least prejudiced distribution for the dice value to determine $\Pr(\text{dice is } \textbf{about\_2} \mid \text{dice is } \textbf{small})$. The mass assignment for **about_2** is given by

$$m_{about\_2} = \{2\} : 0.7, \{1, 2, 3\} : 0.3.$$

We take

$$\begin{aligned} \Pr(\text{dice is } &\textbf{about\_2} \mid \text{dice is } \textbf{small}) \\ &= 0.7\Pr(\text{dice is } 2 \mid \text{dice is } \textbf{small}) + 0.3\Pr(\text{dice is } \{1, 2, 3\} \\ &\qquad\qquad\qquad\qquad\qquad\qquad\qquad\qquad\qquad\quad \mid \text{dice is } \textbf{small}) \\ &= 0.7(0.5333) + 0.3(1) = 0.67331. \end{aligned}$$

In Fril this process of determining the conditional probability is called point semantic unification. Fril can determine $\Pr(\textbf{f}|\textbf{g})$ for any two fuzzy sets **f** and **g**, discrete or continuous, defined on the same domain using this process of point semantic unification. An interval version when no assumption about the prior is also available. A more general form of semantic unification is defined in [12] and a discussion of its properties in [11].

If a derived solution to a query is of the form (answer is **f**) where f is a continuous fuzzy set then a precise value can then be given as the expected value of the least prejudiced density associated with **f**. This corresponds to a defuzzification process but not that normally use in fuzzy control.

3.4   Fuzzy Sets from Data

Consider a rule of the form

((value of y is **g**)
(value of x_1 is **f$_1$**)... (value of x_n is **f$_n$**)) : ((1 1)(0 0)),

and suppose we have an example set of vectors $(y, x_1, \ldots, x_n)$ satisfying the function $y = F(x_1, \ldots, x_n)$. We form a frequency distribution for values of x_i taken from this example set satisfying the head of this rule. We choose a fuzzy set **f$_i$** such that the least prejudiced probability distribution obtained from the mass assignment associated with **f$_i$** is this frequency distribution.

One difficulty we must consider is that the head of the rule contains a fuzzy set **g** so that a given point from the example set will satisfy the head to a certain degree, $\mu_g(y)$. What count do we give to the frequency distribution for x_i for this point? We use the mass assignment theory to answer this question. For the discrete case, let $p_{g^{lp}}$ be the least prejudiced probability distribution found from the mass assignment associated with the discrete fuzzy set **g**. Then we use a count of $p_{g^{lp}}(y)$ for the frequency associated with x_1. Fril treats continuous fuzzy sets **g** in a similar way.

A frequency distribution can therefore be found for x_i for the *k*th rule. The membership function for the fuzzy set **f$_i$** is then chosen such that the least prejudiced probability distribution found from the mass assignment of **f$_i$** is this frequency distribution. This calculation is trivial, as illustrated with the following example.

This can be repeated for each of the variables and fuzzy sets {**f$_i$**} and repeated again for each rule. In treating each of the variables separately we are assuming that the fuzzy set over the product space $F_1 \times \cdots \times F_n$, where F_i is the domain on which **f$_i$** is defined, can be decomposed in this way. This will not always be the case as discussed above, and a more 'intelligent' set of features must be found or redundant features (algebraic combinations of existing ones) introduced. The use of combinations will be discussed in relation to several examples.

Example
Consider variable X which can take values from $\{a, b, c, d\}$ and the frequency distribution of the example set satisfying the given rule is

$$Pr(a) = 0.2, Pr(b) = 0.3, Pr(c) = 0.4, Pr(d) = 0.1.$$

Assume the value of the variable X lies in the fuzzy set **f**.
Let

$$\mathbf{f} = a/\mu_a + b/\mu_b + c/1 + d/\mu_d$$

where $\mu_b \geq \mu_a \geq \mu_d$ so that

$$m_f = c : 1 - \mu_b, \{b, c\} : \mu_b - \mu_a, \{a, b, c\} : \mu_a - \mu_d, \{a, b, c, d\} : \mu_d$$

so that least prejudiced probability distribution is

$$Pr(c) = (1 - \mu_b) + 1/2(\mu_b - \mu_a) + 1/3(\mu_a - \mu_d) + 1/4\mu_d = 0.4$$
$$Pr(b) = 1/2(\mu_b - \mu_a) + 1/3(\mu_a - \mu_d) + 1/4\mu_d = 0.3$$
$$Pr(a) = 1/3(\mu_a - \mu_d) + 1/4\mu_d = 0.2$$
$$Pr(d) = 1/4\mu_d = 0.1$$

so that we require

$$\mu_d = 0.4, \mu_b = 0.9, \mu_a = 0.7$$

giving the fuzzy set

$$\mathbf{f} = a/0.7 + b/0.9 + c/1 + d/0.4.$$

Weights and Fuzzy Sets for Evidential Logic Rule.
The evidential logic rule can take the form

((class k for X)
  (evlog most (
            (feature 1 has value $\mathbf{f}_{k_1}$)w_{k_1}

            $\ldots$

            (feature n has value $\mathbf{f}_{k_n}$)w_{k_n})))
  for $k = 1, \ldots, m$.

Suppose we are given an example set for this classification problem. A frequency distribution over F_i, the domain of $\mathbf{f}_{k_i}$, is determined for feature i for those objects in class k. The fuzzy set $\mathbf{f}_{k_i}$ is then determined by the method described in the previous section. This is repeated for all $\{\mathbf{f}_{k_i}\}$ The near optimal weights in the rules can also be determined by the following heuristic method. This has been found over various problems to give almost optimal weights.

We will consider the choice of weights in rule k. The importance of a given feature, say fr in this rule depends on how well the fuzzy set $\mathbf{f}_{k_r}$ discriminates from the corresponding feature values in the other rules, *i.e.* the greater overlap the fuzzy set $\mathbf{f}_{k_r}$ has with the corresponding fuzzy set in each of the other rules, the less important this feature is in this rule.

We can use point semantic unification to determine the degree of match of $\mathbf{f}_{k_r}$ with $\mathbf{f}_{i_r}$ for all $i, i \neq k$. Let this result in the unifications

$$\mathbf{f}_{k_r}|\mathbf{f}_{i_r} : \theta_{k_{i_r}}; \text{ all } i, i \neq k.$$

The degree of importance of fr in rule k with respect to the ith rule is $1 - \theta_{k_{i_r}}$. The degree of importance of fr depends on $\sum_{i=1; i \neq k}^{m} 1 - \theta_{k_{i_r}}$ so that we define the relative weights of rule k as

$$w'_{k_r} = 1 - \frac{\sum_{i=1;i\neq k}^{m} \theta_{k_{i_r}}}{m - 1} \qquad \text{for r} = 1, \dots, n.$$

The relative set of weights for rule k, namely $\{w'_{k_1}, \dots, w'_{k_n}\}$ are then normalised to give the importance weights for rule k, namely

$$w_{k_i} = \frac{w'_{k_i}}{\sum_{j=1}^{n} w'_{k_j}} \qquad \text{for i} = 1, \dots, n.$$

Filter in Evidential Logic Rule

We have used the fuzzy set 'most' in the evidential logic rule. This can be replaced by any fuzzy set which can give varying interpretations to the rule. For example, if all weights are equal and we use the fuzzy set which has membership 1 only for argument value 1 and zero everywhere else, then we have a hard conjunction. The filter fuzzy set can be chosen to give any interpretation to the rule from a hard conjunction through softer conjunctions to a soft disjunction through to a hard disjunction. By choosing a suitable universe of discourse the fuzzy set can be chosen so that the evidential logic rule is an artificial neuron. In this case we do not restrict the weights to add up to 1 — the weights can have any positive or negative values.

3.5   Simple Example

Consider the following problem of finding a rule to classify the following examples correctly where the three elements of a vector are $(a$ or $\neg a), (b$ or $\neg b)$ and $(c$ or $\neg c)$, respectively. Each vector has the classification (good or bad). Data:

bad vectors

(a b c)
(a ¬b c)
(¬a b ¬c)
(¬a ¬b c)

good vectors

(a b ¬c)
(¬a b c)
(¬a ¬b ¬c)
(a ¬b ¬c).

A suitable rule is:

vector X is good IFF X satisfies $(\neg a \wedge c) \vee (\neg b \wedge \neg c) \vee (a \wedge \neg c)$.

This rule is obtained by deduction since it is true in all possible cases. There are only six possible vectors and these are given the classifications above.

We now modify the problem by giving only the classification of six of the possible eight vectors, namely

Data:

	bad vectors		good vectors
c1.	(a b c)	c2.	(a b $\neg c$)
c4.	(a $\neg b$ c)	c3.	($\neg a$ b c)
c5.	($\neg a$ b $\neg c$)	c6.	($\neg a$ $\neg b$ $\neg c$).

The problem is to find a classification for the vectors $c7 = (\neg a, \neg b, c)$ and $c8 = (a, \neg b, \neg c)$. Of course, there is no correct solution there are four possible classifications for the pair of cards. Both can be good, both can be bad, the first can be good and the second bad, or *vice versa*. We could simply argue that we should use the prior distribution for the classification, namely $\Pr(\text{good}) = \Pr(\text{bad}) = 1/2$. We cannot use Bayes rule, namely

$$\Pr(\text{X is good} \mid \text{vector for X}) = \frac{\Pr(\text{vector for X} \mid \text{X is good})\Pr(\text{X is good})}{\Pr(\text{vector for X})}$$

for the unknown cases since we do not know the required conditional probabilities. Of course, this rule gives 1 or 0 for the known cases dependent on whether the classification is good or bad, respectively. We could make assumptions to determine these conditional probabilities, but how can we make these assumptions? We have, of course, already made an assumption about the prior. Lets make the assumption $\Pr(\text{vector for X} \mid X \text{ is good}) = \Pr(\text{vector for } X)$ for any X then we would obtain $\Pr((\neg a \neg bc) \text{ is good}) = \Pr((a \neg b \neg c) \text{ is good}) = 1/2$.

This assumption ignores the pattern of the vectors. For example, only the bad vectors have just one of the letters present. This may not be relevant, but what about other patterns which belong to one class more frequently than another? We can investigate this sort of discriminatory behaviour in several ways. We will first explore the possibility of finding the simplest rule which gives the correct classification to the known cases, and use this to predict the classifications for the unknown cases. We discussed the use of the simplest rule in the introduction.

ID3 Method

Using ID3 we obtain the decision tree below. This tree gives the classification of good for cach of thc unknown vectors. The decision tree is equivalent to the rule vector X is good IFF X satisfies

$$(\neg a \wedge c) \vee \neg b \wedge c) \vee (a \wedge b \wedge \neg c).$$

Other rules could be found which would give other possible classifications. The rules

vector X is good IF X satisfies $(a \wedge b \wedge \neg c) \vee (\neg a \wedge b \wedge c) \vee (\neg a \wedge b \wedge \neg c)$
vector X is bad IF X satisfies $(a \wedge b \wedge c) \vee (a \wedge \neg b \wedge c) \vee (\neg a \wedge b \wedge \neg c)$

provide no generalization whatsoever, and say nothing about the unknown cases.

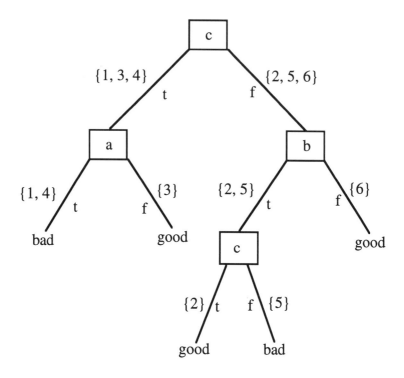

Figure 3.1:

Evidential Logic Rule

We will now explore the use of the Fril evidential logic rule. We will choose all possible feature combinations, namely $A, B, C, A \wedge B, A \wedge C, B \wedge C, A \wedge B \wedge C$. The domain of A is, for example, $\{a, \neg a\}$, the domain of AB is $\{a \wedge b, a \wedge \neg b, \neg a \wedge b, \neg a \wedge \neg b\}$, the domain of $A \wedge B \wedge C$ is $\{a \wedge b \wedge c, a \wedge \neg b \wedge c, a \wedge b \wedge \neg c, a \wedge \neg b \wedge \neg c, \neg a \wedge b \wedge c, \neg a \wedge \neg b \wedge c, \neg a \wedge b \wedge \neg c, \neg a \wedge b \wedge \neg c, \}$, etc. We generate the fuzzy sets for each of these features automatically from the data in a rule for classification good and in the rule for classification bad. The fuzzy sets in these rules are used to determine the importance weights. This gives the evidential logic rules.

Fuzzy Sets

(**aClass1** $\{\neg a : 2/3, a : 1\}$ aDomain)
(**bClass1** $\{\neg b : 2/3, b : 1\}$ bDomain)
(**cClass1** $\{\neg c : 2/3, c : 1\}$ cDomain)
(**abClass1** $\{a \neg b : 1, \neg ab : 1, ab : 1\}$ abDomain)
(**acClass1** $\{\neg a \neg c : 2/3, ac : 1\}$ acDomain)
(**bcClass1** $\{b \neg c : 1, \neg bc : 1, bc : 1\}$ bcDomain)
(**abcClass1** $\{\neg ab \neg c : 1, a \neg bc : 1, abc : 1\}$ abcDomain)

(**aClass2**{¬a : 1, a : 2/3} aDomain)
(**bClass2**{¬b : 2/3, b : 1} bDomain)
(**cClass2**{¬c : 1, c : 2/3} cDomain)
(**abClass2**{¬ab : 1, ¬a¬b : 1, ab : 1} abDomain)
(**acClass2**{a¬c : 1, ¬ac : 1, ¬a¬c : 1} acDomain)
(**bcClass2**{b¬c : 1, ¬b¬c : 1, bc : 1} bcDomain)
(**abcClass2**{¬abc : 1, ¬a¬b¬c : 1, ab¬c : 1} abcDomain).

Rules

((class for X is bad)
  (evlog (
     (value of a for X is **aClass1**) 0.0740741
     (value of b for X is **bClass1**) 0.037037
     (value of c for X is **cClass1**) 0.0740741
     (value of ab for X is **abClass1**) 0.111111
     (value of ac for is **acClass1**) 0.259259
     (value of bc for X is **bcClass1**) 0.111111
     (value of abc for X is **abcClass1**) 0.333333))) : ((1 1) (0 0))

((class for X is good)
  (evlog (
     (value of a for X is **aClass2**) 0.0769231
     (value of b for X is **bClass2**) 0.0384615
     (value of c for X is **cClass2**) 0.0769231
     (value of ab for X is **abClass2**) 0.115385
     (value of ac for X is **acClass2**) 0.230769
     (value of bc for X is **bcClass2**) 0.115385
     (value of abc for X is **abcClass2**) 0.346154))) : ((1 1) (0 0))

These rules for the eight vectors give the following results

(class for c1 is good) : 0.3718   (class for c1 is bad) : 1.0
(class for c2 is good) : 0.7436   (class for c2 is bad) : 0.642
(class for c3 is good) : 0.9744   (class for c3 is bad) : 0.3827
(class for c4 is good) : 0.1282   (class for c4 is bad) : 0.9877
(class for c5 is good) : 0.6538   (class for c5 is bad) : 0.8642
(class for c6 is good) : 0.9872   (class for c6 is bad) : 0.2963
(class for c7 is good) : 0.5   (class for c7 is bad) : 0.2593
(class for c8 is good) : 0.5   (class for c8 is bad) : 0.2593.

Using these results, we classify the vectors as

c1	bad	c2	good	c3	good	c4	bad
c5	bad	c6	good	c7	good	c8	good

The known vectors are classified correctly and both vectors with unknown classifications are classified as good. This is the same result as with ID3.

Simplification of rules

The above rules can be simplified by ignoring features with low importance weights. The weights are recalculated — only the normalization changes — to give

((class for X is bad)
  (evlog (
        (value of ac for X is **acClass1**) 0.259259
        (value of abc for X is **abcClass1**) 0.333333))) :((1 1) (0 0))

((class for X is good)
  (evlog (
        (value of ac for X is **acClass2**) 0.230769
        (value of abc for X is **abcClass2**) 0.346154))) : ((1 1) (0 0))

These rules give the results

(class for c1 is good) : 0	(class for c1 is bad) : 1
(class for c2 is good) : 0.6	(class for c2 is bad) : 0.4375
(class for c3 is good) : 1	(class for c3 is bad) : 0
(class for c4 is good) : 0	(class for c4 is bad) : 1
(class for c5 is good) : 0.4	(class for c5 is bad) : 0.854
(class for c6 is good) : 1	(class for c6 is bad) : 0.292
(class for c7 is good) : 0.4	(class for c7 is bad) : 0
(class for c8 is good) : 0.4	(class for c8 is bad) : 0

giving same classifications as before. The generalization comes from the **acClass1** and **acClass2** fuzzy sets.

Modification of rules

There are other ways of taking the discrimination into account. For example, we can drop terms with the same membership value in the corresponding fuzzy sets in each of the rules to form additional features. These new features will be more highly discriminatory. They should not be put into a separate rule otherwise this could result in an inconsistent rules. Doing this gives the following fuzzy sets and rules:

(**aClass1** {nota:0.666667 a:1})
(**bClass1** {notb:0.666667 b:1})
(**cClass1** {notc:0.666667 c:1})
(**abClass1** {notab:1 ab:1})
(**abNewClass1** {anotb:1})
(**acClass1** {notanotc:0.666667 ac:1})
(**bcClass1** {bnotc:1 bc:1})

(**bcNewClass1** {notbc:1})
(**abcClass1** {notabnotc:1 anotbc:1 abc:1})

(**aClass2** {nota:1 a:0.666667})
(**bClass2** {notb:0.666667 b:1})
(**cClass2** {notc:1 c:0.666667})
(**abClass2** {notab:1 ab:1})
(**abNewClass2** {notanotb:1})
(**acClass2** {anotc:1 notac:1 notanotc:1})
(**bcClass2** {bnotc:1 bc:1})
(**bcNewClass2** {notbnotc:1})
(**abcClass2** {notabc:1 notanotbnotc:1 abnotc:1})

((class for X is bad)
  (evlog (
      (value of a for X is **aClass1**) 0.0513
      (value of b for X is **bClass1**) 0.0256
      (value of c for X is **cClass1**) 0.0513
      (value of ab for X is **abClass1**) 0
      (value of ab for X is **abNewClass1**) 0.2308
      (value of ac for X is **acClass1**) 0.1795
      (value of bc for X is **bcClass1**) 0
      (value of bc for X is **bcNewClass1**) 0.2308
      (value of abc for X is **abcClass1**) 0.2308))) : ((1 1) (0 0))

((class for X is good)
  (evlog (
      (value of a for X is **aClass2**) 0.0526
      (value of b for X is **bClass2**) 0.0263
      (value of c for X is **cClass2**) 0.0526
      (value of ab for X is **abClass2**) 0
      (value of ab for X is **abNewClass2**) 0.2368
      (value of ac for X is **acClass2**) 0.1579
      (value of bc for X is **bcClass2**) 0
      (value of bc for X is **bcNewClass2**) 0.2368
      (value of abc for X is **abcClass2**) 0.2368))) : ((1 1) (0 0))

which gives the results

(class for c1 is good) : 0.0964   (class for c1 is bad) : 0.5385
(class for c2 is good) : 0.3508   (class for c2 is bad) : 0.2906
(class for c3 is good) : 0.5087   (class for c3 is bad) : 0.1111
(class for c4 is good) : 0.0877   (class for c4 is bad) : 0.9916
(class for c5 is good) : 0.2894   (class for c5 is bad) : 0.4445
(class for c6 is good) : 0.9910   (class for c6 is bad) : 0.2051

(class for c7 is good) : 0.4999   (class for c7 is bad) : 0.3334
(class for c8 is good) : 0.4999   (class for c8 is bad) : 0.3334

giving same classifications as before, but as we will see below we would not, in this case, accept this modification.

A combined modified and simplified set of rules are

((class for X is bad)
  (evlog (
        (value of ab for X is **abNewClass1**) 0.2647
        (value of ac for X is **acClass1**) 0.2059
        (value of bc for X is **bcNewClass1**) 0.2647
        (value of abc for X is **abcClass1**) 0.2647))) : ((1 1) (0 0))

((class for X is good)
  (evlog (
        (value of ab for X is **abNewClass2**) 0.2727
        (value of ac for X is **acClass2**) 0.1818
        (value of bc for X is **bcNewClass2**) 0.2727
        (value of abc for X is **abcClass2**) 0.2727))) : ((1 1) (0 0))

giving results

(class for (c1) is good) : 0   (class for (c1) is bad) : 0.4706
(class for (c2) is good) : 0.2727   (class for (c2) is bad) : 0.2059
(class for (c3) is good) : 0.4545   (class for (c3) is bad) : 0
(class for (c4) is good) : 0   (class for (c4) is bad) : 1
(class for (c5) is good) : 0.1818   (class for (c5) is bad) : 0.401967
(class for (c6) is good) : 0.9999   (class for (c6) is bad) : 0.137267
(class for (c7) is good) : 0.4545   (class for (c7) is bad) : 0.2647
(class for (c8) is good) : 0.4545   (class for (c8) is bad) : 0.2647

giving same classifications as before.

The difference in the supports for the decided class minus the support for the other class for the four sets of rules are

	Full	Full Simplified	Modified	Modified Simplified
c1	0.6282	1	0.4421	0.4706
c2	0.1016	0.1625	0.0602	0.0668
c3	0.5917	1	0.3976	0.4545
c4	0.8595	1	0.9039	1
c5	0.2104	0.454	0.1551	0.2202
c6	0.6909	0.708	0.7859	0.8626
c7	0.2407	0.4	0.1665	0.1898
c8	0.2407	0.4	0.1665	0.1898

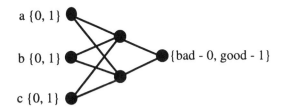

Typical Result

vector	value	decision
c7	0.37	bad
c8	0.37	bad

8556 epochs

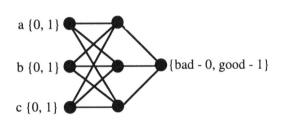

Typical Result

vector	value	decision
c7	0.34	bad
c8	0.38	bad

3185 epochs

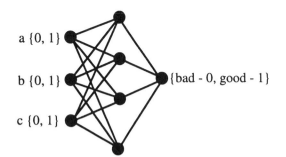

Typical Result

vector	value	decision
c7	0.35	bad
c8	0.52	bad

1812 epochs

Figure 3.2: Neural nets.

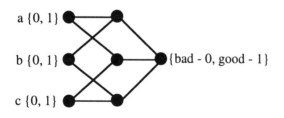

Typical Result

vector	value	decision
c7	0.996	good
c8	0.997	good

very difficult to train
converged mostly to local
minimum with errors

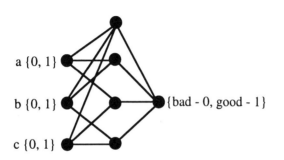

Typical Result

vector	value	decision
c7	0.987	good
c8	0.978	good

Figure 3.3: Neural nets.

From these results we would select the FullSimplified set of rules as our final result. These rules give the best support for the known vector classifications and also provide the highest support for the generalized results.

Neural Nets

Feed forward nets were trained using back-propagation. The results depended very much on the architecture selected and the easy of convergence depended on the connectivity of the nets. In some cases for fully connected nets it was difficult to get zero error.

For this neural net we give solution when training the net from different starting positions:

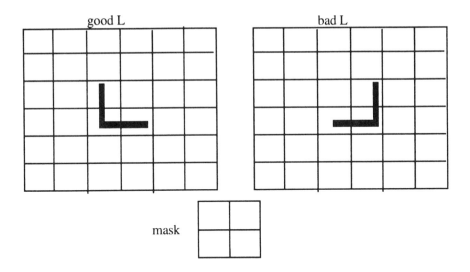

Figure 3.4: Classification of a good L and a bad L.

output		decision	
c7	c8	c7	c8
0.412	0.273	bad	bad
0.585	0.795	good	good
0.991	0.932	good	good
0.930	0.960	good	good
0.956	0.772	good	good
0.841	0.593	good	good
0.987	0.978	good	good
0.220	0.298	bad	bad
0.526	0.970	good	good
0.997	0.996	good	good

The choice of architecture is important, but we have no way of deciding which architecture we prefer by simply considering the neural net. The logic of the situation suggests the last architecture is the one to be preferred. In this case the training of the net mostly results in a {good, good} prediction for the unknown vectors.

3.6   Another Simple Example

Consider the following classification of a good L and bad L.

A mask is randomly dropped onto the grid and moved until some black pixels occur. Patterns for good L and bad L are given below.

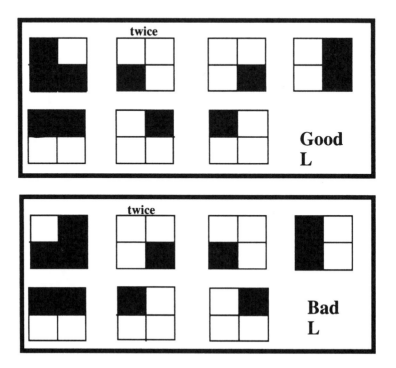

Figure 3.5: Patterns for good L and bad L.

Not all of these patterns are discriminatory. The first and last in each of the top rows are, and the second in each of the top rows occur twice as often as the corresponding pattern in the other box.

It is known that errors in transmission can occur so that the following patterns which are not possible without errors occurring can occur.

We will use the patterns with known classifications to provide a generalized classification for these cases. We use the notation (A, B, C, D) where each element of this vector takes the value 1 if square is black and 0 if square is white, where

A	B
D	C

The data set for learning is:

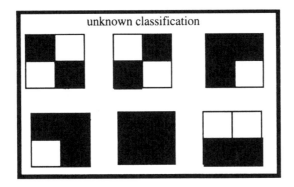

Figure 3.6: Unknown classification.

good L patterns;
(1, 0, 1, 1) ; (0, 0, 0, 1)twice ; (0, 0, 1, 0) ; (0, 1, 1, 0) ; (1, 1, 0, 0) ; (0, 1, 0, 0) ; (1, 0, 0, 0),

bad L patterns;
(0, 1, 1, 1) ; (0, 0, 1, 0)twice ; (0, 0, 0, 1) ; (1, 0, 0, 1) ; (1, 1, 0, 0) ; (1, 0, 0, 0,) ; (0, 1, 0, 0),

unknown classifications patterns;
(1, 0, 1, 0) ; (0, 1, 0, 1) ; (1, 1, 0, 1) ; (1, 1, 1, 0) ; (1, 1, 1, 1) ; (0, 0, 1, 1).

Bayesian Approach

$$\Pr(\text{good L} \mid \text{vector}) = \frac{\Pr(\text{vector} \mid \text{good L})\Pr(\text{good L})}{\Pr(\text{vector}}$$

giving

$\Pr(\text{good L} \mid (1, 0, 1, 1)) = 1$	$\Pr(\text{good L} \mid (0, 0, 1, 0)) = 1/3$
$\Pr(\text{good L} \mid (0, 1, 1, 1)) = 0$	$\Pr(\text{good L} \mid (0, 0, 0, 1)) = 2/3$
$\Pr(\text{good L} \mid (0, 1, 1, 0)) = 1$	$\Pr(\text{good L} \mid (1, 1, 0, 0)) = 1/2$
$\Pr(\text{good L} \mid (1, 0, 0, 1)) = 0$	$\Pr(\text{good L} \mid (0, 1, 0, 0)) = 1/2$
	$\Pr(\text{good L} \mid (1, 0, 0, 0)) = 1/2$

but without a model for how errors can occur we cannot provide any generalization to the unknown classified cases.

For generalization purposes we will assume the following model. Only one transmission error can occur for any mask return and this error is equally likely for all squares. Therefore,

x1 = Pr(good L | (1, 0, 1, 0)) = Pr(good L | (0, 0, 1, 0)Pr(0, 0, 1, 0) +
$\qquad$ Pr(good L | (1, 1, 1, 0)Pr(1, 1, 1, 0) + Pr(good L | (1, 0, 0, 0)
$\qquad$ Pr(1, 0, 0, 0) + Pr(good L | (1, 0, 1, 1)Pr(1, 0, 1, 1)
$\qquad$ = (1/3 + x2 + 1/2 + 1)1/4,

x2 = Pr(good L | (1, 1, 1, 0)) = Pr(good L | (0, 1, 1, 0)Pr(0, 1, 1, 0) +
$\qquad$ Pr(good L | (1, 0, 1, 0)Pr(1, 0, 1, 0) + Pr(good L | (1, 1, 0, 0)
$\qquad$ Pr(1, 1, 0, 0) + Pr(good L | (1, 1, 1, 1)Pr(1, 1, 1, 1)
$\qquad$ = (1 + x1 + 1/2 + x3)1/4,

x3 = Pr(good L | (1, 1, 1, 1)) = Pr(good L | (0, 1, 1, 1)Pr(0, 1, 1, 1) +
$\qquad$ Pr(good L | (1, 0, 1, 1)Pr(1, 0, 1, 1) + Pr(good L | (1, 1, 0, 1)
$\qquad$ Pr(1, 1, 0, 1) + Pr(good L | (1, 1, 1, 0)Pr(1, 1, 1, 0)
$\qquad$ = (0 + 1 + x4 + x2)1/4,

x4 = Pr(good L | (1, 1, 0, 1)) = Pr(good L | (0, 1, 0, 1)Pr(0, 1, 0, 1) +
$\qquad$ Pr(good L | (1, 0, 0, 1)Pr(1, 0, 0, 1) + Pr(good L | (1, 1, 1, 1)
$\qquad$ Pr(1, 1, 1, 1) + Pr(good L | (1, 1, 0, 0)Pr(1, 1, 0, 0)
$\qquad$ = (x5 + 0 + x3 + 1/2)1/4,

x5 = Pr(good L | (1, 1, 0, 1)) = Pr(good L | (1, 1, 0, 1)Pr(1, 1, 0, 1) +
$\qquad$ Pr(good L | (0, 0, 0, 1)Pr(0, 0, 0, 1) + Pr(good L | (0, 1, 1, 1)
$\qquad$ Pr(0, 1, 1, 1) + Pr(good L | (0, 1, 0, 0)Pr(0, 1, 0, 0)
$\qquad$ = (x4 + 2/3 + 0 + 1/2),

so that

4x1 = x2 + 11/6
4x2 = x1 + x3 + 3/2
4x3 = x2 + x4 + 1
4x4 = x3 + x5 + 1/2
4x5 = x4 + 7/6,

giving

x1 = 28/45, x2 = 59 / 90, x3 = 1/2, x4 = 31/90, x5 = 17/45.

Also

Pr(good L | (0, 0, 1, 1)) = Pr(good L | (1, 0, 1, 1)Pr(1, 0, 1, 1) +
$\qquad$ Pr(good L | (0, 1, 1, 1)Pr(0, 1, 1, 1) + Pr(good L | (0, 0, 0, 1)
$\qquad$ Pr(0, 0, 0, 1) + Pr(good L | (0, 0, 1, 0)Pr(0, 0, 1, 0)
$\qquad$ = (1 + 0 + 2/3 + 1/3)1/4 = 1/2.

Thus the generalized Bayesian solution using our error model assumption is

Pr(good L | (1, 0, 1, 0) = 0.6222 ; Pr(good L | (1, 1, 1, 0) = 0.6556 ;
Pr(good L | (1, 1, 1, 1) = 0.5 ; Pr(good L | (1, 1, 0, 1) = 0.3444
Pr(good L | (0, 1, 0, 1) = 0.3778 ; Pr(good L | (0, 0, 1, 1) = 0.5.

Our Bayesian generalizations are therefore

good L patterns:
(1, 0, 1, 0) ; (1, 1, 1, 0),
bad L patterns:
(1, 1, 0, 1) ; (0, 1, 0, 1),
uncertain patterns:
(1, 1, 1, 1) ; (0, 0, 1, 1).

Evidential Logic Rule

The Fril evidential logic program automatically generated from the learning data set is given below. The notation is A = {a,¬ a} where a corresponds to black square for A and a to white square. Similar interpretation for B, C and D.

/* Fuzzy sets for body features */
(**abcdClass1** {notabcd:0.875 abnotcnotd:0.875 notanotbcnotd:1
notanotbnotcd:0.875 notabnotcnotd:0.875 anotbnotcd:0.875
anotbnotcnotd:0.875})

(**bcdClass1** {notbnotcnotd:0.625 bcd:0.625 notbcnotd:1 bnotcnotd:1
notbnotcd:1})

(**acdClass1** {notacd:0.75 notanotcnotd:0.75 notacnotd:1 anotcd:0.75
anotcnotd:1 notanotcd:0.75})

(**abdClass1** {notabd:0.875 notanotbnotd:1 notabnotd:0.875 anotbd:0.875
abnotd:0.875 anotbnotd:0.875 notanotbd:0.875})

(**abcClass1** {notabc:0.75 notanotbnotc:0.75 notabnotc:0.75 abnotc:0.75
anotbnotc:1 notanotbc:1})

(**abcdClass2** {abnotcnotd:0.875 notanotbcnotd:0.875 notanotbnotcd:1
anotbcd:0.875 notabcnotd:0.875 notabnotcnotd:0.875 anotbnotcnotd:0.875})

(**bcdClass2** {notbcd:0.75 bcnotd:0.75 notbnotcnotd:0.75 notbcnotd:0.75
bnotcnotd:1 notbnotcd:1})

(**acdClass2** {notanotcnotd:0.625 notacnotd:1 anotcnotd:1 notanotcd:1
acd:0.625})

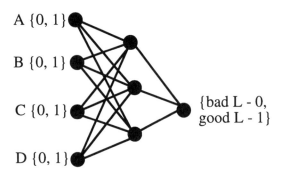

Figure 3.7: The feed forward net.

(**abdClass2** {notanotbnotd:0.75 notabnotd:1 anotbd:0.75 abnotd:0.75
anotbnotd:0.75 notanotbd:1})

(**abcClass2** {notabc:0.875 notanotbnotc:1 notabnotc:0.875 anotbc:0.875
abnotc:0.875 anotbnotc:0.875 notanotbc:0.875}).

/* Rules */

((bad support for X)
     (evlog (
          (abcd of X in **abcdClass1**) 0.265823
          (bcd of X in **bcdClass1**) 0.240506
          (acd of X in **acdClass1**) 0.177215
          (abd of X in **abdClass1**) 0.0886076
          (abc of X in **abcClass1**) 0.227848))) : ((1 1) (0 0))

((good support for X)
     (evlog (
          (abcd of X in **abcdClass2**) 0.265823
          (bcd of X in **bcdClass2**) 0.177215
          (acd of X in **acdClass2**) 0.240506
          (abd of X in **abdClass2**) 0.227848
          (abc of X in **abcClass2**) 0.0886076))) : ((1 1) (0 0)).

For these rules we have used only the combinations ABC and ABCD. We
could have also used the combinations AB, AC, AD, BC, BD, CD and also the
separate attributes A, B, C and D. We have used the combination of three to
give a result which we can compare with the Bayesian results which assumed
that only one error could occur. The results using these additional features
do not change very much in support values, and the decisions are as given

below.

A simplified set of rules can also be used, and these will give similar results without any change in the decisions. This also applies to a modified set of rules using the same modification procedure as used in the previous example.

If we apply these rules to the complete data set we obtain the results:

Pattern	Support for Good L	Support for Bad L	Decision
Learning Set			
1011	0.7642	0.0775	good L
0111	0.0775	0.7642	bad L
0110	0.9114	0.4256	good L
1001	0.4256	0.9114	bad L
0010	0.8544	1.0000	bad L
0001	1.0000	0.8544	good L
1100	0.8987	0.8987	uncertain
0100	0.8665	0.8544	uncertain
1000	0.8544	0.8655	uncertain
Generalization			
1010	0.3813	0.3180	good L
0101	0.3181	0.3813	bad L
1101	0.0775	0.3038	bad L
1110	0.3038	0.0775	good L
1111	0.1503	0.1503	uncertain
0011	0.4383	0.4383	uncertain

The learning sets are correctly predicted. We use 0.05 as the threshold for making a decision with respect to the difference in the supports. This really applies to the generalization set where we do not expect such large differences because of the nature of the evidential logic rule which contains the feature ABCD with important weight which cannot be satisfied by the unknown classified patterns. We can say that we have high confidence in the decisions for the generalizations for patterns $(1, 1, 0, 1)$, $(1, 1, 1, 0)$, $(1, 1, 1, 1)$ and $(0, 0, 1, 1)$. We are less confident about the decisions for patterns $(1, 0, 1, 0)$ and $(0, 1, 0, 1)$ but if forced to make a decision then the ones given are those we would make.

The decision results are the same as for the Bayesian solution with the given error model.

Neural Net Solutions

Using back-propagation with the feed forward net (see Figure 3.7) gives the following results:

Pattern	Neural Net Output − >0.5 for good L ; <0.5 for bad L	Decision
Learning Set		
1011	0.883	good L
0111	0.099	bad L
0110	0.912	good L
1001	0.086	bad L
0010	0.316	bad L
0001	0.688	good L
1100	0.503	uncertain
0100	0.529	uncertain
1000	0.487	uncertain
Generalization		
1010	0.478	uncertain
0101	0.080	bad L
1101	0.069	bad L
1110	0.916	good L
1111	0.099	bad L
0011	0.882	good L

For the net (Figure 3.8), we get the results:

Pattern	Neural Net Output − >0.5 for good L ; <0.5 for bad L	Decision
Learning Set		
1011	0.913	good L
0111	0.107	bad L
0110	0.916	good L
1001	0.090	bad L
0010	0.298	bad L
0001	0.677	good L
1100	0.502	uncertain
0100	0.511	uncertain
1000	0.488	uncertain
Generalization		
1010	0.921	good L
0101	0.527	uncertain
1101	0.088	bad L
1110	0.927	good L
1111	0.913	good L
0011	0.108	bad L

These results do not agree with the Bayesian or evidential logic solution.

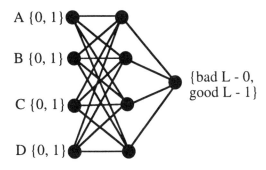

Figure 3.8:

The two neural nets both give good results for the learning set since the outputs are close to the true Pr(good L | vector), but they give contradictory generalizations. The patterns (1, 1, 1, 1) and (0, 0, 1, 1) are predicted as bad L and good L, respectively, for the first neural net while for the second net the predictions are good L and bad L, respectively. Both these differ from the Bayesian and evidential logic predictions which are that both patterns give uncertainty in the classifications. There are other discrepancies between the neural net and evidential logic results.

ID3 method

The decision tree provided using this method is can be seen in Figure 3.9. This decision tree gives the following results

Pattern	ID3 Pr(good L | vector)	Decision
Learning Set		
1011	1	good L
0111	0	bad L
0110	1	good L
1001	0	bad L
0010	1/3	bad L
0001	2/3	good L
1100	1/2	uncertain
0100	1/2	uncertain
1000	1/2	uncertain
Generalization		
1010	1/2	uncertain
0101	1/2	uncertain
1101	1/2	uncertain
1110	1/2	uncertain
1111	1/2	uncertain
0011	1/2	uncertain

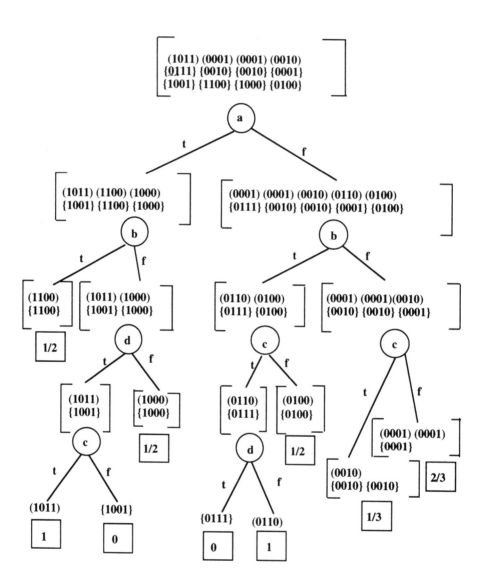

Figure 3.9: Decision tree

The Bayesian probabilities are produced for the training set but the tree has no power of generalization. This, of course, is understandable since in the absence of any error model or other means of matching sub-patterns the best that we can do is to predict uncertain.

3.7 A Golf Problem

The following database has been used to illustrate C4.5, an extension of ID3:

outlook	temp	humidity	windy	Class
sunny	85	85	false	Don't Play
sunny	80	90	true	Don't Play
overcast	83	78	false	Play
rain	70	96	false	Play
rain	68	80	false	Play
rain	65	70	true	Don't Play
overcast	64	65	true	Play
sunny	72	95	false	Don't Play
sunny	69	70	false	Play
rain	75	80	false	Play
sunny	75	70	true	Play
overcast	72	90	true	Play
overcast	81	75	false	Play
rain	71	80	true	Don't Play

The domain for the attribute outlook is {sunny, overcast, rain}, that for windy is {true, false}. The domains for temperature and humidity are ranges of values, namely [60, 90] and [60, 100], respectively. The domain for Class is {Play, Don't Play}. We wish to find a rule so that we can predict the Class when given information concerning the other attributes of the database. Suppose we wish to predict the class when the outlook is sunny, temperature is 85, humidity is 80 and windy is false.

Evidential Logic Solution

The Fril evidential logic rule for this example gives the result Don't Play. It uses the combination of attributes outlook and humidity, i.e the cross product space OUTLOOK × HUMIDITY. This takes the form of (sunny × **f1**) + (overcast × **f2**) + (rainy × **f3**) where **f1, f2, f3** are fuzzy sets on the humidity domain. **f1, f1'** are the fuzzy sets for HUMIDITY when outlook attribute is sunny, **f2, f2'** are the fuzzy sets for HUMIDITY when the outlook is overcast and **f3, f3'** are the fuzzy sets for HUMIDITY when the outlook is rainy. Fril can construct fuzzy sets from a few points, but uses assumptions which the user will be asked to accept. For example the fuzzy set **f1** is given by Figure 3.10.

Only one point is given for Class Play and three points for Class Don't

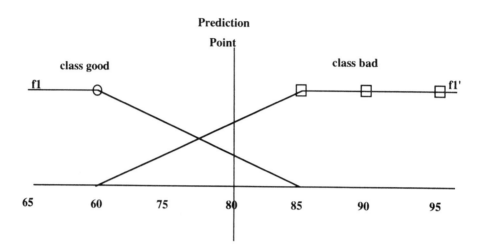

Figure 3.10:

Play. The fuzzy sets given are reasonable for this data and can be used to predict the class. The user will be told that this assumption was used and can accept or modify the fuzzy sets.

C4.5 solution

The decision tree is in Figure 3.11 which predicts the same solution for the given case as Fril, namely Don't Play. This result is dependent on the way the continuous attributes are chunked into discrete attributes.

Neural Net

The following provides the relevant information for a back propagation feed forward neural net (see Figure 3.12).

This neural net gives zero error for the learning data set provided by the database, but does not provide an intuitive pleasing result for the given case requiring generalization. Again it illustrates the danger of using a neural net for generalization.

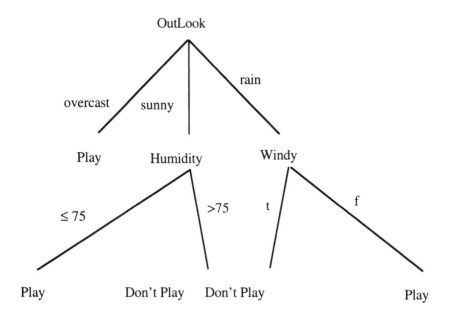

Figure 3.11: The decision tree

3.8 Other Examples

The data browser has been applied to more practical examples. The recognition of under water sounds in the presence of noise were discussed in [8]. The automated Fril methods without taking combinations were applied and gave 92% accuracy on the test set. The test set was different to the training set. 32 features were used for this problem.

It has also been applied to hand written character recognition on a restricted set of difficult letters — {a, o, u, l, d}. A large database of badly written letters was used to find the fuzzy sets and weights in the evidential logic rule. Unintelligent features which were simply number of black pixels in strips of the pixel set were used and the accuracy obtained was of the order of 89%. More intelligent features were used using additional evidential logic rules for the recognition of various curve shapes in each of four regions. This can give better results but requires the use of combinations of features. This work will be reported later.

The evidential logic rule method gave 94% on the MIT face data which uses 18 face attributes to recognize whether a face belongs to a male or female. This 94% can be improved using a modification of the fuzzy sets as discussed in

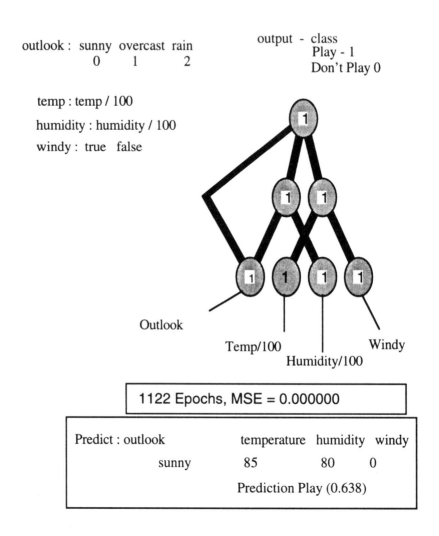

outlook : sunny overcast rain
              0   1   2

output - class
        Play - 1
        Don't Play 0

temp : temp / 100

humidity : humidity / 100

windy : true false

Outlook

Temp/100   Windy

Humidity/100

1122 Epochs, MSE = 0.000000

Predict : outlook   temperature humidity windy
              sunny   85   80   0

        Prediction Play (0.638)

Figure 3.12:

the first example above and using certain combinations. Genetic programming can be used to find appropriate combinations.

Identifying legal points [9], ellipse, compared with illegal points lying outside the ellipse has also been discussed. The Fril method used the X and Y co-ordinates as features but also constructed additional features, namely $X - Y$ and $X + Y$ to overcome the decomposition error. The fuzzy sets for these four features were constructed automatically from a data set of randomly chosen points and a fuzzy control type rule used for the recognition.

This gave 92% accuracy with slight variation dependent on the orientation of the ellipse with respect to the axes.

3.9 A Neuro Fuzzy Method

In this section we will use the fuzzy sets extracted from the data to provide inputs to a neural net. The actual inputs to the neural net are the semantic unification of the example feature value with the corresponding feature fuzzy set for each of the classifications. The neural net will be used instead of the evidential logic rule. We will explain the method by applying it to the card and the L examples previously discussed. The results for the card example will be found to be consistent with the evidential logic method and intuitively plausible. For the L example this method provides good generalization in most cases. For one new case the generalization is not intuitive and differs from that given by the evidential logic rule.

Card Example

The fuzzy sets for each of the three features for good card and bad card were found previously and are

good:
$$f_{ga} = a/2/3 + \neg a/1$$
$$f_{gb} = b/1 + \neg b/2/3$$
$$f_{gc} = c/2/3 + \neg c/1$$

bad:
$$f_{ba} = a/1 + \neg a/2/3$$
$$f_{bb} = b/1 + \neg b/2/3$$
$$f_{bc} = c/1 + \neg c/2/3$$

These six fuzzy sets are used as input features to the neural net. The value of a feature F with fuzzy set f is given by the semantic unification $Pr(f|$ value of feature from example). Thus the input vectors, $(Pr(f_{ga}|\text{data}), Pr(f_{gb}|\text{data}),$ $Pr(f_{gc}|\text{data}), Pr(f_{ba}|\text{data}), Pr(f_{bb}|\text{data}), Pr(f_{bc}|\text{data})$ for the example set are:

good:
$$(2/3, 1, 1, 1, 1, 2/3)$$
$$(1, 1, 2/3, 2/3, 1, 1)$$
$$(1, 2/3, 1, 2/3, 2/3, 2/3)$$

bad:
$$(2/3, 1, 2/3, 1, 1, 1)$$
$$(2/3, 2/3, 2/3, 1, 2/3, 1)$$
$$(1, 1, 1, 2/3, 1, 2/3)$$

We use the neural net (see Figure 3.13).

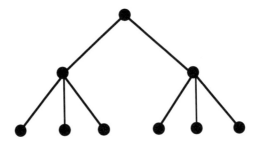

Figure 3.13: The neural net

Training using back propagation gives the following results:

Original data
good: Pr(card is good)
   (2/3, 1, 1, 1, 1, 2/3)   0.935
   (1, 1, 2/3, 2/3, 1, 1)   0.936
   (1, 2/3, 1, 2/3, 2/3, 2/3) 0.974

bad:

   (2/3, 1, 2/3, 1, 1, 1)   0.024
   (2/3, 2/3, 2/3, 1, 2/3, 1) 0.024
   (1, 1, 1, 2/3, 1, 2/3)   0.099

New data
  (1, 2/3, 2/3, 2/3, 2/3, 1) 0.994 c7 = (a, b, c)
  (2/3, 2/3, 1, 1, 2/3, 2/3) 0.994 c8 = (a, b, c)

This result is consistent with that using the evidential logic rule.

L example

If we use the features A, B, C, D then the fuzzy sets for good L and bad L are

good L:
$$f_{ga} = a/3/4 + \neg a/1$$
$$f_{gb} = b/3/4 + \neg b/1$$
$$f_{gc} = c/3/4 + \neg c/1$$
$$f_{gd} = d/3/4 + \neg d/1$$

bad:

$$f_{ba} = a/3/4 + \neg a/1$$
$$f_{bb} = b/3/4 + \neg b/1$$
$$f_{bc} = c/3/4 + \neg c/1$$
$$f_{bd} = d/3/4 + \neg d/1$$

Since these are all the same they will not be useful.

We therefore go to the next level of combination and use a combination of two features. The combinations are AB, AC, AD, BC, BD, CD. The fuzzy sets extracted from the data example set are:

good:

$$f_{gAB} = ab/0.5 + a\neg b/0.875 + \neg ab/0.875 + \neg a\neg b/1$$
$$f_{gAC} = ac/0.5 + a\neg c/0.875 + \neg ac/0.875 + \neg a\neg c/1$$
$$f_{gAD} = ad/0.5 + a\neg d/0.875 + \neg ad/0.875 + \neg a\neg d/1$$
$$f_{gBC} = bc/0.5 + b\neg c/0.875 + \neg bc/0.875 + \neg b\neg c/1$$
$$f_{gBD} = bd/0 + b\neg d/1 + \neg bd/1 + \neg b\neg d/0.75$$
$$f_{gCD} = cd/0.5 + c\neg d/0.875 + \neg cd/0.875 + \neg c\neg d/1$$

bad:

$$f_{bAB} = ab/0.5 + a\neg b/0.875 + \neg ab/0.875 + \neg a\neg b/1$$
$$f_{bAC} = ac/0 + a\neg c/1 + \neg ac/1 + \neg a\neg c/0.75$$
$$f_{bAD} = ad/0.5 + a\neg d/0.875 + \neg ad/0.875 + \neg a\neg d/1$$
$$f_{bBC} = bc/0.5 + b\neg c/0.875 + \neg bc/0.875 + \neg b\neg c/1$$
$$f_{bBD} = bd/0.5 + b\neg d/0.875 + \neg bd/0.875 + \neg b\neg d/1$$
$$f_{bCD} = cd/0.5 + c\neg d/0.875 + \neg cd/0.875 + \neg c\neg d/1$$

The vector input sets,

(Pr(faAB | data), Pr(faAC | data), Pr(faAD | data), Pr(faBC | data), Pr(faBD | data), Pr(faCD | data), Pr(fbAB | data), Pr(fbAC | data), Pr(fbAD | data), Pr(fbBC | data), Pr(fbBD | data), Pr(fbCD | data))

for the neural net are therefore

good:
(0.875, 0.5, 0.5, 0.875, 1, 0.5, 0.875, 0, 0.5, 0.875, 0.875, 0.5)
(1, 1, 0.875, 1, 1, 0.875, 1, 0.75, 0.875, 1, 0.875, 0.875) twice
(1, 0.875, 1, 0.875, 0.75, 0.875, 1, 1, 1, 0.875, 1, 0.875)
(0.875, 0.875, 1, 0.5, 1, 0.875, 0.875, 1, 1, 0.5, 0.875, 0.875)
(0.5, 0.875, 0.875, 0.875, 1, 1, 0.5, 1, 0.875, 0.875, 0.875, 1)
(0.875, 1, 1, 0.875, 1, 1, 0.875, 0.75, 1, 0.875, 1)
(0.875, 0.875, 0.875, 1, 0.75, 1, 0.875, 1, 0.875, 1, 1, 1)

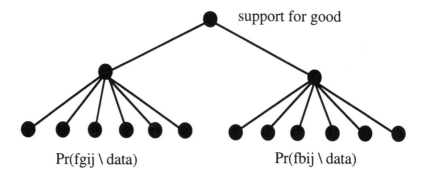

Figure 3.14:

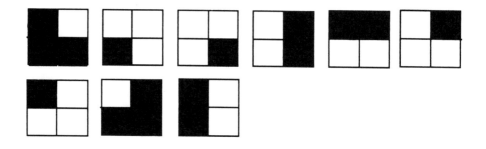

Figure 3.15:

bad:
(0.875, 0.875, 0.875, 0.5, 0, 0.5, 0.875, 1, 0.875, 0.5, 0.5, 0.5)
(0.875, 0.875, 0.5, 1, 1, 0.875, 0.875, 1, 0.5, 1, 0.875, 0.875)
(1, 1, 0.875, 1, 1, 0.875, 1, 0.75, 0.875, 1, 0.875, 0.875)
(1, 0.875, 1, 0.875, 0.75, 0.875, 1, 1, 1, 0.875, 1, 0.875) twice
(0.5, 0.875, 0.875, 0.875, 1, 1, 0.5, 1, 0.875, 0.875, 0.875, 1)
(0.875, 1, 1, 0.875, 1, 1, 0.875, 0.75, 1, 0.875, 1)
(0.875, 0.875, 0.875, 1, 0.75, 1, 0.875, 1, 0.875, 1, 1, 1)

We use the neural net (see Figure 3.14).

The results for the input data vectors are:

Figure 3.16:

	Support for good L
(0.875, 0.5, 0.5, 0.875, 1, 0.5, 0.875, 0, 0.5, 0.875, 0.875, 0.5)	0.982
(1, 1, 0.875, 1, 1, 0.875, 1, 0.75, 0.875, 1, 0.875, 0.875)	0.603
(1, 0.875, 1, 0.875, 0.75, 0.875, 1, 1, 1, 0.875, 1, 0.875)	0.357
(0.875, 0.875, 1, 0.5, 1, 0.875, 0.875, 1, 1, 0.5, 0.875, 0.875)	0.953
(0.5, 0.875, 0.875, 0.875, 1, 1, 0.5, 1, 0.875, 0.875, 0.875, 1)	0.502
(0.875, 1, 1, 0.875, 1, 1, 0.875, 0.75, 1, 0.875, 1)	0.549
(0.875, 0.875, 0.875, 1, 0.75, 1, 0.875, 1, 0.875, 1, 1, 1)	0.444
(0.875, 0.875, 0.875, 0.5, 0, 0.5, 0.875, 1, 0.875, 0.5, 0.5, 0.5)	0.105
(0.875, 0.875, 0.5, 1, 1, 0.875, 0.875, 1, 0.5, 1, 0.875, 0.875)	0.030

These results are for Figure 3.15, respectively.

The results for the new data vectors are:

	support for good L
(0.875, 0.5, 0.875, 0.875, 0.75, 0.875, 0.875, 0, 0.875, 0.875, 1, 0.875)	1.000
(0.875, 1, 0.875, 0.875, 0, 0.875, 0.875, 0.75, 0.875, 0.875, 0.5, 0.875)	0.000
(0.5, 0.875, 0.5, 0.875, 0, , 0.875, 0.5, 1, 0.5, 0.875, 0.5, 0.875)	0.000
(0.5, 0.5, 0.875, 0.5, 1, 0.875, 0.5, 0, 0.875, 0.5, 0.875, 0.875)	1.000
(0.5, 0.5, 0.5, 0.5, 0, 0.5, 0.5, 0, 0.5, 0.5, 0.5, 0.5)	<u>0.932</u>
(1, 0.875, 0.875, 0.875, 1, 0.5, 1, 1, 0.875, 0.875, 0.875, 0.5)	0.481

These data vectors correspond to Figure 3.16, respectively.

The underlined result does not agree with intuition. The intuitive result is 0.5. This arises because of over fitting of the data. If we control this over fitting by stopping the iteration when the error for intuitive result starts to increase we get the results.

The results for the input data vectors when controlling over fitting are:

	Support for good L
(0.875, 0.5, 0.5, 0.875, 1, 0.5, 0.875, 0, 0.5, 0.875, 0.875, 0.5)	0.936
(1, 1, 0.875, 1, 1, 0.875, 1, 0.75, 0.875, 1, 0.875, 0.875)	0.528
(1, 0.875, 1, 0.875, 0.75, 0.875, 1, 1, 1, 0.875, 1, 0.875)	0.497
(0.875, 0.875, 1, 0.5, 1, 0.875, 0.875, 1, 1, 0.5, 0.875, 0.875)	0.688
(0.5, 0.875, 0.875, 0.875, 1, 1, 0.5, 1, 0.875, 0.875, 0.875, 1)	0.512
(0.875, 1, 1, 0.875, 1, 1, 0.875, 0.75, 1, 0.875, 0.875, 1)	0.601
(0.875, 0.875, 0.875, 1, 0.75, 1, 0.875, 1, 0.875, 1, 1, 1)	0.399
(0.875, 0.875, 0.875, 0.5, 0, 0.5, 0.875, 1, 0.875, 0.5, 0.5, 0.5)	0.054
(0.875, 0.875, 0.5, 1, 1, 0.875, 0.875, 1, 0.5, 1, 0.875, 0.875)	0.156

and the results for the new data vectors when controlling over fitting are:

	support for good L
(0.875, 0.5, 0.875, 0.875, 0.75, 0.875, 0.875, 0, 0.875, 0.875, 1, 0.875)	0.984
(0.875, 1, 0.875, 0.875, 0, 0.875, 0.875, 0.75, 0.875, 0.875, 0.5, 0.875)	0.016
(0.5, 0.875, 0.5, 0.875, 0, , 0.875, 0.5, 1, 0.5, 0.875, 0.5, 0.875)	0.004
(0.5, 0.5, 0.875, 0.5, 1, 0.875, 0.5, 0, 0.875, 0.5, 0.875, 0.875)	0.994
(0.5, 0.5, 0.5, 0.5, 0, 0.5, 0.5, 0, 0.5, 0.5, 0.5, 0.5)	0.429
(1, 0.875, 0.875, 0.875, 1, 0.5, 1, 1, 0.875, 0.875, 0.875, 0.5)	0.515

The generalization results are now good but the original data is not that well satisfied. In fact if the neural net is trained on the whole data, *i.e.* training + test data, the results are not that good. We will therefore use fuzzy sets corresponding to combinations of three features as our inputs to the neural net. This compares with what we did for the evidential logic rule. The fuzzy sets used here are the same as for the evidential logic rule.

We now use combinations of three, namely

$$ABC, \ ACD, \ ABD, \ BCD$$

using this neuro-fuzzy approach.

The fuzzy sets for this case are:

good L:

$$f_{gABC} = \text{abc} \ / \ 0 + \text{ab c} \ / \ 0.875 + \text{a bc} \ / \ 0.875 + \text{a b c} \ / \ 0.875 + \\ \text{abc} \ / \ 0.875 + \text{ab c} \ / \ 0.875 + \text{a bc} \ / \ 0.875 + \text{a b c} \ / \ 1$$

$$f_{gACD} = \text{acd} \ / \ 0.625 + \text{ac d} \ / \ 0 + \text{a cd} \ / \ 0 + \text{a c d} \ / \ 1 + \text{acd} \ / \ 0 + \\ \text{ac d} \ / \ 1 + \text{a cd} \ / \ 1 + \text{a c d} \ / \ 0.625$$

$$f_{gABD} = \text{abd} \ / \ 0 + \text{ab d} \ / \ 0.75 + \text{a bd} \ / \ 0.75 + \text{a b d} \ / \ 0.75 + \text{abd} \ / \ 0 + \\ \text{ab d} \ / \ 1 + \text{a bd} \ / \ 1 + \text{a b d} \ / \ 0.75$$

$$f_{gBCD} = \text{bcd} \ / \ 0 + \text{bc d} \ / \ 0.75 + \text{b cd} \ / \ 0 + \text{b c d} \ / \ 1 + \text{bcd} \ / \ 0.75 + \\ \text{bc d} \ / \ 0.75 + \text{b cd} \ / \ 1 + \text{b c d} \ / \ 0.75$$

bad L:
$$f_{bABC} = abc \; / \; 0 + ab \; c \; / \; 0.75 + a \; bc \; / \; 0 + a \; b \; c \; / \; 1 + a\overline{b}c \; / \; 0.75 + \\ a\overline{b} \; c \; / \; 0.75 + a \; \overline{b}c \; / \; 1 + a \; \overline{b} \; c \; / \; 0.75$$

$$f_{bACD} = acd \; / \; 0 + ac \; d \; / \; 0 + a \; cd \; / \; 0.75 + a \; c \; d \; / \; 1 + a\overline{c}d \; / \; 0.75 + \\ a\overline{c} \; d \; / \; 1 + a \; \overline{c}d \; / \; 0.75 + a \; \overline{c} \; d \; / \; 0.75$$

$$f_{bABD} = abd \; / \; 0 + ab \; d \; / \; 0.875 + a \; bd \; / \; 0.875 + a \; b \; d \; / \; 0.875 + \\ a\overline{b}d \; / \; 0.875 + a\overline{b} \; d \; / \; 0.875 + a \; \overline{b}d \; / \; 0.875 + a \; \overline{b} \; d \; / \; 1$$

$$f_{bBCD} = bcd \; / \; 0625 + bc \; d \; / \; 0 + b \; cd \; / \; 0 + b \; c \; d \; / \; 1 + b\overline{c}d \; / \; 0 + \\ b\overline{c} \; d \; / \; 1 + b \; \overline{c}d \; / \; 1 + b \; \overline{c} \; d \; / \; 0.625$$

These provide the input vectors:

good L:
(0.875, 0.625, 0.75, 0.75, 0, 0, 0.875, 0)
(1, 1, 1, 1, 0.75, 0.75, 0.875, 1) twice
(0.875, 1, 0.75, 0.75, 1, 1, 1, 1)
(0.875, 1, 1, 0.75, 0.75, 1, 0.875, 0)
(0.875, 1, 0.75, 1, 0.75, 1, 0.875, 1)
(0.875, 1, 1, 1, 0.75, 1, 0.875, 1)
(0.875, 1, 0.75, 0.75, 1, 1, 0.875, 0.625)

bad L:
(0.875, 0, 0, 0, 0.75, 0.75, 0.875, 0.625)
(0.875, 1, 0.75, 0.75, 1, 1, 1, 1) twice
(1, 1, 1, 1, 0.75, 0.75, 0.875, 1)
(0.875, 0, 0.75, 1, 1, 0.75, 0.875, 1)
(0.875, 1, 0.75, 1, 0.75, 1, 0.875, 1)
(0.875, 1, 0.75, 0.75, 1, 1, 0.875, 0.625)
(0.875, 1, 1, 1, 0.75, 1, 0.875, 1)

The neural net is in Figure 3.17, these results for the data set are in Figure 3.18, and the results for the new set are in Figure 3.19.

Evidential Logic Rules for Neuro-Fuzzy Net

The nets obtained by the neuro-fuzzy method can be represented by Fril evidential logic rules. The above net for the L problem would be represented by three Fril Evidential Logic Rules — one rule consisting of features with the fuzzy sets for good L, namely $\{f_{gABC}, \dots\}$ and one rule with features containing fuzzy sets for bad L, namely $\{f_{bABC}, \dots\}$. The weights in these rules are those neural weights given in the net. The filter fuzzy set in these evidential logic rules is the activation function of the neural net. The third evidential logic rule is one containing the heads of the previous evidential logic rule. The weights for this rule are chosen from the neural net above and

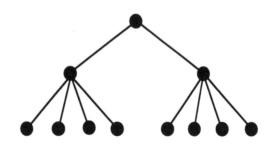

Figure 3.17:

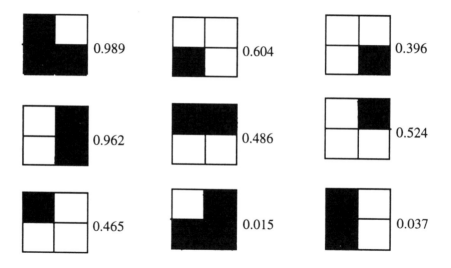

Figure 3.18:

again the filter fuzzy set id as before. Each evidential logic rule can take into account the biases by introducing an added feature which is always satisfied.

3.10   Conclusions

A general method, which uses fuzzy sets and importance weights in an evidential logic rule or just fuzzy sets in a fuzzy control type rule, for induction has been described. Combination of features were used to provide better generalization. This use of combinations requires further work when there are many possible combinations to consider. Genetic programming can be used

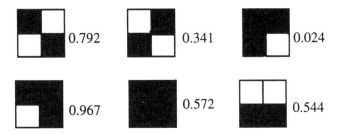

Figure 3.19:

to select appropriate combinations. The feature domains can be a truth set {t, f}, a more general discrete set or a range of values. Combinations can use features with different types of domains. The examples given in this chapter discuss problems requiring the use of combinations.

The examples give a comparison of the Fril evidential logic method with the use of ID3 and its extension C4.5 and also the use of neural nets. The Fril method performed well on all examples, while the other methods had certain difficulties. A comparison was also made with a Bayesian approach.

A neuro-fuzzy approach was also discussed, and the results showed promise when the over fitting error was controlled. For practical problems this over fitting error is controlled by monitoring the error in the test set along side the error in the training set as the back propagation iteration proceeds. The iteration is stopped when the test set error starts to rise. The resulting neural net should provide good generalization to unseen cases.

References

[1] Baldwin J.F. (1991) Approximate Reasoning, Fuzzy & Probabilistic Control using a theory of Mass Assignments. *Proc. of IFES* Japan, 611–622.

[2] Baldwin J.F. (1992) Fuzzy and Probabilistic Uncertainties. In *Encyclopaedia of AI* 2nd edition, Shapiro (ed), Wiley, 528–537.

[3] Baldwin J.F. (1992) Evidential reasoning under Probabilistic and Fuzzy Uncertainties. In *An Introduction to Fuzzy Logic Applications in Intelligent Systems* Yager R.R. and Zadeh L.A. (eds), Dordrecht: Kluwer, 297–333.

[4] Baldwin J.F. (1993) A Mass Assignment Theory and Memory Based Reasoning. In *Intelligent Systems with Uncertainty* Bouchon-Meunier, Valverde and Yager (eds), Elsevier, 97–107.

[5] Baldwin J. F. (1993) Fuzzy Reasoning in Fril for Fuzzy Control and other Knowledge Based Applications. *Asia-Pacific Engineering Journal* (Part A) 3 (Parts 1 & 2): 59–81.

[6] Baldwin J.F. (1993) Fuzzy Sets, Fuzzy Clustering and Fuzzy rules in AI. *Fuzzy Logic in AI Workshop, 13th Int. Joint Conf. on Artificial Intelligence* Chambery, France, 1–11.

[7] Baldwin J.F. (1993) Evidential Support Logic, Fril and Case based Reasoning. *Int J. of Intelligent Systems* 939–960.

[8] Baldwin J.F., Gooch R.M. and Martin T.P. (1994) Classification of Underwater Sounds using Evidential Logic. *Proc. EUFIT 94* (1) 106–110.

[9] Baldwin J.F. and Martin T.P. (1995) Fuzzy Modelling in an Intelligent Data Browser. *Proc. of IEEE / FUZZ95* 1885–1890.

[10] Baldwin J.F. and Pilsworth B.W. (1995) Proc. Adaptive Decision technolgies 95, Unicom Seminars.

[11] Baldwin J.F., Lawry J. and Martin T.P. (To Appear in Fuzzy Sets and Systems), A mass Assignmnet Theory of the Probability of Fuzzy Events.

[12] Baldwin J.F., Martin T.P. and Pilsworth B.W. (1995) *FRIL — Fuzzy and Evidential Reasoning in AI.* Research studies Press & Wiley.

[13] Dubois D. and Prade H. (1982) On several representations of an uncertain body of evidence. In *Fuzzy Information and Decision Processes* Gupta M.M. and Sanchez E. (des), North Holland Pub. Co.

[14] Dubois D., Prade H. and Testemale C. (1986) Fuzzy pattern matching with extended capabilities: Proximity notions, importance assessment, random sets. *Proceedings of NAFIPS'86 — Recent Developments in the Theory and Applications of Fuzzy Sets.*

[15] Fril Systems Ltd. (1986) *Fril Manual.*

[16] Nguyen H.T. (1978) On random sets and belief functions. *J. Math. Anal. & Appl.* 65: 531–542.

[17] Shafer G. (1976) *A Mathematical Theory of Evidence.* Princeton Univ. Press.

[18] Smets Ph. and Kennes R. (1989) The transferable belief model: Comparison with Bayesian Models. Iridia Tech. Report.

[19] Zadeh L.A. (1965) Fuzzy Sets. *Information and Control* 8: 338–353.

[20] Zadeh L.A. (1978) PRUF — A meaning representation language for natural languages. *Int, J. Man-Machine Studies* 10: 395–460.

4

Mission Management System for Multiple Autonomous Vehicles

N.J.W Rayner, C.J Harris

4.1 Introduction

This chapter discusses the engineering of Mission Management Knowledge Based processes for the command of Multiple Intelligent Autonomous Vehicles (MIAVs). In particular, it discusses architectural and algorithmic considerations in the light of the demanding requirements for robustness and increased system longevity. The architectural issues covered reflect recent developments in Object Technology, which has demonstrated the benefits of a componentised view of systems, where observation of interface standards can provide for a 'plug and play' approach to system development and evolution. The algorithmic considerations concentrate on the significant progress in the machine learning field specifically looking at combining popular Knowledge Based Systems (KBSs) approaches with those developed in the area of Artificial Neural Networks (ANNs). Adaptive systems promise resistance to change through the modification of internal models as a result of direct experience of the problem domain, and can, under certain conditions, behave robustly in unseen situations.

The engineering of Mission Management Knowledge Based processes for the command of Multiple Intelligent Autonomous Vehicles (MIAVs) concerns the coordination of elements of a distributed system so as to generate coherent behaviour. As such the techniques apply to the management and control of envisaged civil information and automation systems in public utilities, transportation and manufacturing as well as military command and control. The

ever increasing demand for cost effectiveness, project efficiency and increasing
productivity resulting from open competition is resulting in greater demand
for coherent, systems solutions for bespoke large scale projects, such as major
building construction, air traffic control and road management systems. These
integrated systems are characterized by high capital value and extended life
time, which together raise a requirement for evolution in system capability
and the acceptance of change as an inherent characteristic of system infrastructure. The acceptance of change implies the need for a rigorous approach
to the analysis and design of these systems which emphasizes the achievement
of modularity. In addition, since these systems often are required to operate
in dynamic large, complex, uncertain, unstructured, non-benign environments
without human intervention, there is a requirement for an intelligent adaptive
ability which can react to environmental dynamics. Adaptive systems are
more resistant to system and environmental changes potentially resulting in
significant cost saving though increased operational life.

The engineering of Mission Management systems for Multiple Intelligent
Autonomous Vehicles (MIAVs) in particular and the management problem
domain in general places heavy reliance on human decision making and supervision. Computerized management systems have been difficult to introduce
primarily as a result of inadequacies in the technology. This is, in part, due
to difficulties with describing models of the domain with sufficient precision.
Experts in management have a good 'feel' for problems in the domain but,
despite being effective managers, find it difficult to express their knowledge in
anything but an approximate, vague, rule of thumb way. This conflicts with
computer systems requirements which need a precise and complete description
of domain relationships.

Robotic computerized management solutions traditionally involve the
Knowledge Based Planning (KBP) of activities over time and their monitoring
during execution. This research extends KBP to approximate rule based systems to support initialization from expert knowledge, while supporting adaption to fine tune approximate rule sets to better describe domain relationships.
This approach takes advantage of expressible human expertise while compensating for inaccuracy, ignorance and incompleteness by supporting adaption,
giving systems increased resistance to change and therefore greater longevity.
Advantage is taken of recent developments in the use of approximate rule
based models in the initialization of adaptive control algorithms, specifically
Neurofuzzy algorithms which can tune an approximate model to reflect arbitrary process relationships [9]. Neurofuzzy Networks not only have the well
understood adaptive advantages of Associative Memory Networks, but also
can be initialized with symbolic fuzzy linguistic rules improving their transparency, and thus aiding in system development and maintainability.

While neurofuzzy networks are an extremely effective tool in nonlinear
modelling [9], they suffer from the *curse of dimensionality* [39]; consequently
the size of neurofuzzy systems, and the required computing resources, grow exponentially with respect to the input space dimension. This renders the use of

neurofuzzy systems on high-dimensional problems impractical; management problems typically have hundreds of variables from numerous separate subject domains which makes the naive adoption of neurofuzzy techniques problematic. The limitations of the neurofuzzy approach are a result of assuming total ignorance on the part of the designer which allows no structure to be imposed on networks to reduce their size. Structure imposes limitations on the generality of a network such that it cannot reflect arbitrary process relationships, however imposition of structure offers the possibility, in practise, of solving complex real world problems. Two ways in which structure can be imposed are; the introduction of expert and/or designer knowledge which can be used to separate out network substructures which are specific to certain variable relationships; and automated construction where statistical analysis can be used to identify and impose structure.

For a knowledge based planner to build plans it must have a representation of the knowledge associated with performing actions in the real world. These actions are represented by operators which define the causal relationship between an actions conditions of initiation and the resultant effects of its execution. The causal relationship must be able to describe not only the immediate effects of action execution, but its future consequences as well. The causal effects of an action relate to the context in which it is executed including its relative position within a plan. To capture the property of action sequence, causal models are built across *cases* (sequences of actions) which are equivalent to plans. The adoption of Case Based Planning (CBP) helps to enhance the reactivity of the planning system reducing effort spent in search and plan construction by storing plan sequences for reapplication in similar situations. The storing of cases(plans) for future use requires the management of the finite computing resource, this is achieved through the pruning of cases. Cases that are similar are compounded to form generalized cases which are applicable to a wider range of problems. In addition, significant relationships are abstracted out of the cases through an analysis of variance on the causal models. This results in the generation of an abstract case structure with wide applicability and the freeing of some of the storage consumed by the original cases.

The architecture of the MIAV system takes advantage of recent developments in component oriented systems which have been a byproduct of the Object Technology movement. The architecture supports the distribution of vehicles and vehicle components across separate platforms, supporting a 'plug and play' metaphor for the iterative development and maintenance of the system. For demonstration purposes, the vehicles interact in a simulated game where two teams are in opposition, and the vehicles cooperate both constructively and destructively in order to exercise various cooperative strategies.

This work has, in part, been based on research undertaken in the Advanced Systems Research Group (now the Image, Speech and Intelligent Systems Research Group) on an ESPRIT II project PANORAMA (Perception And Navigation fOR Autonomous Mobile Applications) which ended in October 1993. The principle integration testbed was a Mercedes 4-wheel-drive vehicle REMI; additional smaller laboratory based robots were used for local system integration and testing, whilst the developed final demonstration was performed on a tracked drilling machine, owned by TAMROCK (Finland). The drilling machine was required to accurately navigate through an unstructured environment, the intention being to perform drilling operations at various drill sites with a location accuracy of $+/-$ 5cm. The 10MECU EU Esprit II CIM project represented a major component in the EU research strategy to address automation problems in its industrial base. This chapter describes a project called PSYCHE which involves the ongoing implementation of a Task Level Mission Management system for cooperating Intelligent Autonomous Vehicles (IAVs) (funded by the ESPRC).

In section 4.2 Case Based Planning is reviewed followed by an introduction to fuzziness (section 4.3) and its representation as part of a neurofuzzy implementation. The adaption of causal models is described in section 4.4, and techniques for addressing the *curse of dimensionality* are discussed in the following section 4.5. Finally, the concluding sections describe the architecture and implementation of the PSYCHE MIAV system.

4.2   Case Based Planning

In knowledge based planning a variety of techniques have been developed and instantiated in different systems. A 'state of the art' classical planner would normally adopt a mixture of these techniques (the following are of some of the more important features):

- Search and 'Means-End Analysis' (STRIPS [14, 15, 16]).

- Hierarchical search spaces (ABSTRIPS [34]).

- Nonlinear Representations (NONLIN [33], NOAH[12]).

- Critics (NOAH [12], NONLIN[33]).

- Causal Models (SIPE [41]).

Case based planning is probably the most promising development in knowledge based planning in recent times. It takes as a starting premise that the organization of experience is paramount in formulating new plans and debugging old ones. The CBP framework suggested by Hammond [18] has six basic CBP processes:

- An **Anticipator** that predicts planning problems on the basis of failures experienced in similar situations.

- A **Retriever** that searches a plan memory for a plan which satisfies as many of the current goals as possible while avoiding the problems that the anticipator has predicted.

- A **Modifier** that alters the plan found by the retriever to achieve any goals from the input that it does not satisfy.

- A **Storer** that places new plans in memory, indexed by the goals that they satisfy and the problems they avoid.

- A **Repairer** that is called if a plan fails, and which then repairs plans and records causal explanations for plan failure.

- An **Assigner** that uses the causal explanation built during repair to determine the features which will predict this failure in the future. This knowledge is used to index the failure for later anticipation. As in repair, causal knowledge is useful in anticipation [18].

Traditional CBP techniques suffer, as other KBSs do, with brittleness. To address this concern many have tried to integrate neural network technology into KBSs [38]. Ram *et al.* [31], for example, propose a self-improving navigation system that uses a multistrategy learning method to allow the system to adapt to novel environments, which combines reinforcement with CBP. Difficulties have existed, however, with trying to retain the expressive power of symbolic representations utilized in traditional KBSs. Recent developments using approximate symbolic relationships, rather than precise ones, have appeared, where expertise can be expressed in a comparatively transparent form. The enabling techniques are in the field of fuzzy logic which can be used to help describe the vagueness associated with relations in the world, but still retain some of the important linguistic properties.

4.3 Fuzziness

Fuzziness is concerned with the lack of well-defined boundaries of the set of objects to which some symbol refers, reflecting linguistic imprecision, uncertainty as well as process complexity [20].

A fuzzy set $\tilde{A}$ in a universe of discourse Ω is characterized by the function:

$$\mu_{\tilde{A}}(x):\Omega \rightarrow [0,1] \tag{4.1}$$

So, for example, a fuzzy subset $F\tilde{A}ST$ of a universe of discourse 'speeds of robot'. The function $\mu_{\tilde{A}}(x)$ returns 'grades of membership' of the fuzzy subset in the range 0 to 1 (0 being not a member, 1 complete membership).

The 'grade of membership' can have different interpretations. Zadeh [42] refers to $\mu_{\tilde{A}}(x)$ as the 'degree of possibility' that x is the value fuzzily restricted by $\tilde{A}$. The fuzzy subset $F\tilde{A}ST$, for example, describes the 'speeds of robot' that could possibly be described as fast and to what degree. Baldwin [6] refers to fuzzy sets as distributions across probability spaces. There has been intense and sometimes bitter discussion concerning interpretation of fuzzy sets and their relationship to the various forms of uncertainty reasoning evidenced in human problem solving and linguistic expression. To date there is no concensus as to the correct interpretation. Here the interpretation is the more general relational or set-theoretic interpretation. Where the 'grades of membership' reflect an ordering of the objects in the universe induced by, in our example, $F\tilde{A}ST$ on 'speeds of robot'. Precise membership values do not exist by themselves they are tendency indices that are subjectively assigned by an individual or group [13].

The above approach initiated by Zadeh provides a tool for modelling human-centered systems [43]. This is because it supports linguistic variables which allow complex situations to be summarized in a manner sympathetic to a human interpretation.

Fuzzy rules are often referred to as *vague* (implying non-specific or approximate relationships) and because they are expressed by humans they are frequently subjective, context dependent and behave as summaries and generalizations of actual relationships. Experiences gained in the practice of Knowledge Elicitation have demonstrated the great difficulty associated with the human ability to explain expertise. It would seem that much expertise becomes transformed into automatic reactive response making it difficult to recall, and therefore making it hard for the expert to be precise.

Computer solutions to problems depend on precise specifications or at least any specification that is not precise becomes so when implemented on a computer:

> *"The imprecision associated with a set of fuzzy rules is generally completely resolved once they have been implemented on a computer as specific meanings have been assigned to the vague statements."* [29].

There is then often a translation process where natural language gets mapped into something computer friendly. A useful distinction is made here between a *fuzzy algorithm* which contains a set of imprecise, qualitative, linguistic rules, and a *fuzzy system* which refers to a specific implementation of the rules Procyk and Mamdani [30]. The *fuzzy algorithm* then is those rules expressed in a linguistic manner extracted from an expert. Once the fuzzy rules are *translated* into a specification or implemented on a computer, they loose their fuzzy/vague nature becoming what is termed a *fuzzy system*.

What also characterizes the 'expert knowledge' input into the system is its generalized nature in that it tries to catch all contexts associated with a relationship. So although the rules may have general applicability they may not necessarily be very effective in all contexts. The human expert, may well subconsciously recognize these different contexts and modify his behaviour accordingly. The computer implementation does not have access to this information, but what it can do is modify its behaviour with respect to direct experience, therefore increasing its context sensitivity. In general:

> *"The power of a fuzzy approach lies in the way these vague imprecise rules can be given a specific meaning and the way the*

fuzzy system produces outputs for inputs which only partially match the rules used in initialization or training." [9]

4.3.1 Neurofuzzy Interpretation

Fuzzy systems, like neural networks, perform a nonlinear input-output mapping (discrete or continuous), but with a set of linguistic rules. A typical rule might be as follows:

IF error is *small* AND change in error is *large*
THEN output is *small* (c_{ij})

where error and change in error are the inputs, *small* and *large* are fuzzy linguistic terms, and (c_{ij}) is a rule confidence. A fuzzy system consists of the union (OR) of a set of these rules, from which a fuzzy output is inferred. The vague linguistic variables are represented by precise fuzzy membership functions, the form of which determine the system's modelling capabilities. The choice of the fuzzy membership functions and the underlying fuzzy logic operators are left to the discretion of the modeller. It has been shown [9] that if the membership functions are B-splines, and the product operator is used to represent the fuzzy intersection (AND) and implication (IF $\cdots$ THEN), and the addition operator is used to represent the fuzzy union (OR), the fuzzy system is equivalent to a B-spline network. The antecedent of each rule represents each multidimensional basis function $(a_i(\mathbf{x}))$; the intersection of a set of linguistic variables defined on the different input variables. The consequence and the rule confidences (c_{ij}) can be represented by weights (w_i), where a direct invertible relationship governs this equivalence. The fuzzy output sets must be chosen as symmetrical B-splines and typically the standard seven triangular linguistic sets are chosen. The neurofuzzy rule based approach is explained in detail in [9].

4.3.2 Curse of Dimensionality

The number of rules in a neurofuzzy system is given by:

$$\text{number of fuzzy rules} \propto \prod_{i=1}^{n}(p_i),$$

where $i = (1, \ldots, n)$ and p_i represents the number of linguistic variables (univariate B-splines) defined on each input dimension. It can be seen that this number increases exponentially with the number of input variables (n). This dilemma is common to many modelling techniques and is termed the *curse of dimensionality*. As a result, conventional high-dimensional neurofuzzy systems are impractical, where the size of the system, and hence the memory requirements and the size of the teaching set required for a given resolution, grows exponentially with respect to the input space dimension. To make the use of high-dimensional neurofuzzy systems feasible the number of fuzzy rules must be reduced without adversely affecting the quality of the approximation. This can be achieved by construction algorithms which, from a teaching set, construct parsimonious neurofuzzy models in both low and high dimensions [8]. Often conventional model structures produce many redundant rules and hence alternative representations can be used to provide similar behaviour but with increased simplification. The resulting parsimonious models should possess the following properties:

- better generalization;

- a reduction in the amount of training data required;

- a better insight into the structure and behaviour of the modelled process;

- simpler rules; and

- a reduction in the computational cost.

Traditionally neurofuzzy model identification is performed in an *ad hoc* manner by encoding expert knowledge, in the form of fuzzy rules, into the neurofuzzy model. The weights of this neurofuzzy model can then be adjusted to manipulate the nonlinear mapping. This approach relies on the availability of sufficient expert knowledge in the form of fuzzy rules, describing the relationship being modelled. Such knowledge is often limited or unavailable, which coupled with the curse of dimensionality has motivated the development of construction algorithms [8]. The process of construction is usually an iterative process of statistical analysis and performance measurement followed by some change in the model structure. Here we are interested in the pruning of the causal models associated with actions in plans as well as plans themselves.

4.4 Adaptive Planning

4.4.1 Operators and Causal Models

For a knowledge based planner to build plans it must have a representation of the knowledge associated with performing actions in the real world. These actions are represented by operators which define the causal relationship between an actions conditions of initiation and the resultant effects of its execution. The causal relationships are captured within causal models which allow context dependent effects to be inferred. Wilkins describes causal models for operators with two distinct sets of rules; state rules and causal rules [41]. State rules are used to infer new conditions about a particular state, while causal rules are used to infer the effects of executing a particular action. A set of causal or state rules relate some set of conditions to some new set of conditions and can be viewed as an input/output mapping which can be represented by a set of fuzzy rules. Adaption of operators is performed by modifying state rules and causal rules to reflect the mismatch between the model and the real world. Since a set of causal or state rules which relate inputs to some output can be considered to be a fuzzy subnetwork. An implicit surface is generated relating input fuzzy sets to output fuzzy sets. As described earlier, these rules can be given confidence ranges in the interval $[0, 1]$. By modifying the confidences across parts of a fuzzy subnetwork one can emphasise or deemphasize certain rules meaning that the output shifts to reflect this change of emphasis. In this sense, the rules can be said to adapt (learn). To generate a sensible change in confidences and produce the desired learning effects there is a requirement for the development of a performance measure. The performance measure relates the difference between what was expected and what actually happened (the error) and the rate that that difference changes over time (change in error) to some change in the confidence value.

$$\text{If } \triangle\text{error and error Then } \triangle\text{confidence.} \qquad (4.2)$$

Given a set of these rules to cover the possible set of $\triangle$errors and errors one can build a rule set to change the rule confidences of the causal and state fuzzy subnetworks (rules sets). Since the representation of the performance measure is rule based, it is interesting to consider adapting this in turn, the idea then being that the system may be able to learn how to learn (for processes more analytical in nature, direct modification of the rule base confidence vector can be achieved by LMS type learning rules [9]). For the purposes of illustration, consider a military scenario where a vehicle has a choice of whether to *approach* or *run* when confronted with an enemy vehicle given that it is near base (and relative safety) and has a degree of health ranging from strong to weak. A simple fuzzy causal rule base may be of the following form (the real numbers indicate the confidences in each fuzzy rule and strong, approach_quickly, etc. are labels of fuzzy sets):

IF strong AND near_base THEN approach_quickly 1.0
OR IF strong AND near_base THEN stop_and_wait 0.0
OR IF strong AND near_base THEN retire_quickly 0.0

OR IF ok AND near_base THEN approach_quickly 0.8
OR IF ok AND near_base THEN stop_and_wait 0.2
OR IF ok AND near_base THEN retire_quickly 0.0

OR IF weak AND near_base THEN approach_quickly 0.0
OR IF weak AND near_base THEN stop_and_wait 0.2
OR IF weak AND near_base THEN retire_quickly 0.8

By examining the confidence values, it can be seen that the fuzzy rule set is weighted towards the approach behaviour implying that the vehicle can be more aggressive when near its base, since it can retire quickly to relative safety should the need arise. The adaption is controlled by a fuzzy learning rule set which defines the learning function.

IF die AND killed_enemy THEN punish_strongly 0.1
OR die AND killed_enemy THEN punish_weakly 0.8
OR die AND killed_enemy THEN no_change 0.1

. .

. .

OR IF greatly_weakened AND killed_enemy THEN no_change 0.8
OR IF slightly_weakened AND killed_enemy THEN encourage_weakly 0.8
OR IF no_weakening AND killed_enemy THEN encourage_strongly 0.8

. .

etc.

The learning rule set punishes activities that result in loss of own life, while encouraging activities that result in enemy kills. The learning strategy can be made quite complicated, since the description of the learning function in terms of linguistic rules tends to improve transparency. Different learning strategies may be adopted for different causal models or domains; for example, one causal model may have a learning strategy that seeks the best solution (but takes time), while another may seek only reasonable performance (but is fast).

4.5   Strategic Knowledge and Planning

A strategy is usually a recurring sequence of actions which addresses a problem given a particular context. As a planner interacts with the world, it is confronted with a stream of goals, or sets of conjunctive goals, rather than independent problems. If these are all treated singly, independently, the total planning and execution effort would be at least the sum of the separate costs. If the planner can learn or represent specific plans that are tailored to a domain of recurring sets of goals, it can avoid much computational effort inherent in weak methods. In other words, being able to 'remember' frequently used and successful sequences can greatly reduce computational complexity [18]. Strategies are accessed by the set of goals they satisfy and the current state of the world. A strategy has its *own* set of fuzzy subnetworks defining the causal relationships between input (preconditions) and output (postconditions). The results of executing a strategy are not necessarily the same as would be deduced through inferring the causal behaviour from its primitive components. This is because the primitive operators are general purpose which allows them to be utilized in a variety of different situations. A strategy is specific to achieving a sequence of goals, and therefore can be seen as

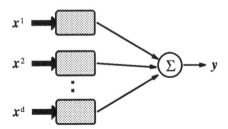

Figure 4.1: General ANOVA decomposition, consisting of d submodels, where $\mathbf{x^i}$ represents a small subset of the input variables ($\mathbf{x^i} \subset \mathbf{x}$).

specialized to that sequence of activities. When a primitive operator adapts, it adapts to make it generally more suitable to a number of contexts, while a strategy is tuned to a comparatively specific problem. Strategies, importantly, reflect an emergence of specialized skills.

4.5.1 Merging Strategic Knowledge

It is necessary to compress strategic knowledge where possible to support continuous behaviour within limited computational resources. Without an active pruning process the quantity of stored strategic knowledge would exceed the space required to store it. An approach is to merge similar causal components shared by different strategies, this can be done by an analysis of the strategies' neurofuzzy causal models. Conventional neurofuzzy systems are *complete*; every input forms part of the antecedent of each rule. In such a representation certain combinations of input variables, *i.e.* rules maybe redundant, and are hence not required to reproduce the desired function. In this case the model can be *globally partitioned,* for example, the output can be represented by the ANalysis Of VAriance (ANOVA) decomposition:

$$f(\mathbf{x}) = f_0 + \sum f_i(x_i) + \sum f_{i,j}(x_i, x_j) + \cdots + \sum f_{1,2,\ldots,n}(\mathbf{x})$$

where $f(\mathbf{x})$ is an additive composition of subfunctions of smaller dimensions. For many functions this complete representation is unnecessary as certain input variable combinations are redundant, and hence can be omitted from the approximation, making a less complex and more parsimonious model. When applied to conventional neurofuzzy systems individual components (subfunctions) of the ANOVA representation can be modelled by a conventional lattice neurofuzzy model. This gives a neurofuzzy system composed of a series of simpler submodels, the outputs of which are summed (ORed) to produce the overall model output, as illustrated in Figure 4.1. The resulting rules within the submodels have antecedents involving a smaller subset of the input vector, making them simpler and more interpretable. Also, as the output of each subnetwork is linear in the weights the individual weight vectors can be concatenated and normal neurofuzzy adaption rules can be applied. Neurofuzzy models representing the causal behaviour of strategies can be compared by comparing the common subfunction components, where subfunctions are sufficiently similar duplication of the information is unrequired and space can be saved.

4.5.2   Abstracting Strategic Knowledge

Abstract strategic knowledge is the result of the abstraction of sequences of operators (including other abstractions) which map some input (initial state) to an output (goal state) that characterize the abstraction. In other words, the input/output relationships represented in the abstraction are typically the most significant relationships that differentiate one higher level of abstraction from another lower level. Abstract knowledge was used firstly in ABSTRIPS [32], where it was shown to simplify the planning process and increase computational efficiency. The process of automatic abstraction is performed by identifying the most significant subfunction components in the ANOVA representation. Those subfunctions that contribute little to the causal relationship are considered to be less significant and exist lower in the abstraction hierarchy. (A previous UNICOM book covered numerically driven algorithms such as adaptive spline modelling (ASMOD) to produce additive models [39].)

The employment of hierarchical representations supports interleavement of planning and execution. Interleavement encourages the production of plans which contain sufficient detail to achieve near term goals while outlining intentions for long term goals. This allows the planner to determine the success of actions, before committing itself to computationally expensive detailed planning for future tasks which may be wasted due to some near term plan failure. The interleavement of planning and execution, therefore, reduces the computational time consumed in planning by reducing unwarranted commitment to plans. Using interleavement enhances reactivity by postponing planning until immediate problems are dealt with.

4.6   System Architecture and Implementation

Object-oriented development is the process of systems development using an *Object* based abstraction of system structure and composition. A system is modelled in terms of the objects and interactions within the problem domain, with both real world and conceptual entities being represented by abstract data types. Object-oriented development is one element of the wider Object Technology (OT) movement, which encompasses object-oriented programming languages, object-oriented analysis and design, object databases, object based graphical user interface design and distributed object management systems. OT represents a movement towards the development of systems from components, where the location and implementation of a given component is transparent to other system elements, enabling the development of open, distributed systems.

Thus, Object Technology can be applied to the construction of IAV Systems at two fundamental levels:

- at the component level, for the production of system software that provides system functionality, *and*

- at the infrastructural level, as a model for system structure and for construction from components, whether built using objects or not.

Object orientation promises open systems, and implies the interoperability of components and the portability of components and applications across heterogeneous platforms and operating systems, with ease of integration of their separate components into coherent systems. It represents the design of **interchangeable components within distributed systems**, for which a number of principles can be defined:

Composability is the ease of construction of total systems from structured collections of components,

Transparency is the independence of the interoperation of two components from their physical locations and implementations,

Extensibility is the ease of extension of a total system, by extension to existing components or addition of new components,

Scalability is the evolution of a system from the small to the large scale, representing increasingly complex abstractions and wider geographical extent,

Portability is the ease of implementation of a system, or system component, on multiple physical architectures (platforms),

Compatibility is the ease with which total system components can be replaced or interchanged,

Interoperability is the ease with which total system components can communicate and interoperate to provide *system level* functionality.

In software engineering, object orientation represents a component based approach to system synthesis, and is being increasingly used to tackle the problems associated with software complexity and distributed systems. The object model views system entities as components within the total system, these components being defined by a strong interface that encapsulates the implementation of that component. The object model includes mechanisms for representing commonality between components and supporting reuse of

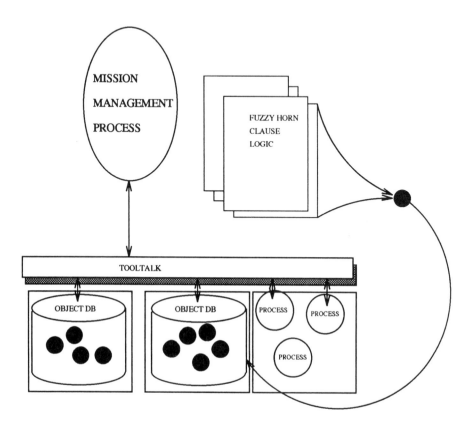

Figure 4.2: Object-oriented implementation of an agent.

existing component resources. Interaction between components is modelled by the passing of messages, the protocols for these messages being defined by the interfaces presented by individual and related groups of components. This model has proven to be very suitable for modelling distributed systems, where the aim is for 'off-the-shelf' components which can be incorporated within new systems without modification, each contributing to *total* system functionality. This is sometimes referred to as a 'plug and play' metaphor.

The object-oriented programming language C++ was selected for the project after a comparative study during project definition. A number of options were considered, including Lisp, PROLOG, CLOS, Eiffel and KEE, as well as C++. Object orientation could provide greater readability, maintainability and extensibility of code, through modularity based on the encapsulation of data and procedures with information hiding. C++ also provided for strong type checking at compile time, and with no run-time garbage collection and access to memory management facilities, computational performance could be reasonable. (NB. It is worth noting that C++ is not an easy language to use and requires a good six months experience.)

The advantage of using an object-oriented programming language for system development is that the language can be evolved such that it better models the problem domain. The representational power of the language is extended by the addition of user classes and objects, with the effect that the implementation becomes less and less involved with primitive C++, and more involved with the objects and associated methods that are components in the evolving model. This inherent extensibility of an object based model is a principle factor in the achievement of quality in implemented systems, though it should be noted that simply using an object-oriented programming language will not assure quality. Rather, this is dependent on the use of development frameworks and infrastructures that support the construction of systems from components. To accommodate the desired infrastructure the Tooltalk interface tool was utilized since it imposes a communication standard that supports the construction of interoperable system components. The Tooltalk standard precedes the increasingly more popular CORBA compliant ORBs (Object Request Brokers) which are promising to impose wide software communication standards encouraging the creation of componentised modular systems.

To exercise the PSYCHE planning system a multi-vehicle combat simulation/gaming environment has been adopted. The program — Xtank (which is public domain) — supports a rich array of agent behaviours and agent designs. Vehicles can be equipped with armour, weapons, fuel and many other features. Numbers of scenarios can be executed, with two or more vehicles, acting constructively as a group, or destructively as members of competing groups. Agents may communicate (via Tooltalk) in a variety of ways (see Figure 4.3);

- Broadcast — to all agents.

- Private — between two agents.

- Group — to members of a team only.

- Individual — team members to one agent.

Xtank is a multiagent simulator with a very rich set of possible configurations; very simple or complex scenarios may be played out to test various management strategies. In addition, being public domain, expertise exists in it's use, most importantly in the development of Xtank agents. Xtank, therefore, is potentially useful for the comparison of independent agent designs.

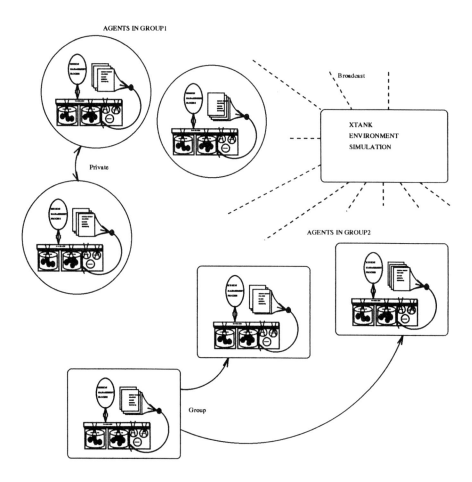

Figure 4.3: Multi-agent architecture communicating through Tooltalk.

4.7   MIAV Mission Management System Architecture

The architecture of the Fuzzy Planner follows the Case Based Planning paradigm. With reference to Figure 4.4, the principle components of the architecture are the Meta-Controller, Planner and Adapter.

4.7.1   The Meta-Controller

The Meta-Controller process has overall control of all subsystem processes. This process functions as the planning system manager. Its main function is to coordinate planning and learning activities. The coordination of these activities is context dependent, in other words, when time is short, the adaption process may be postponed enabling the Planner to react in a timely fashion.

4.7.2 Planner Architecture

The Planner generates plans given information concerning the current state, and the set of goals and their constraints. It searches the knowledgebase for information that will allow the IAV to reach its goals and maintain the generated plan. Actions are passed out to other processes in the IAV architecture for execution. The internals of the planner are depicted in Figure 4.5. The submodules of the planning process are as follows:

- **A Retriever**. The Retriever searches for primitive actions/operators or partial plans (strategies) that satisfy as many of the current goals as possible. The search is constrained by the current set of goals and associated constraints, and also the conditions within the current state.

- **A Modifier**. This process alters the plan found by the retriever to achieve any goals from the input that it does not satisfy. This may well involve generating new subgoals which the Retriever will be requested to find possible solutions to.

- **A Monitor**. The process of monitoring the actions and their execution is performed by the Monitor submodule. This process compares the expected results of performing an action to the actual. In this way the Planner identifies plan failures.

- **A Repairer**. The Repairer is called if a plan fails. It attempts to patch the existing plan. If it is unsuccessful it requests the Modifier to restart.

4.7.3 Adapter Architecture

The Adapter process modifies the knowledge base to reflect experience from acting out actions and plans in the real world. It receives inputs from the new state of the world, the intended action that generated the new state, and also the plan of which this action was a part. The internals of the Adapter are depicted in Figure 4.6. The submodules of the adapting process are as follows:

- **Rule Modifier**. The Rule Modifier adapts causal and state rules. The process of adaption is through the emphasis and de-emphasis of rules through the increment and decrement of rule confidences (see section 4.4).

- **Plan Storer**. The Storer places new plans in the knowledge base; these represent the strategies previously described. The plans are indexed by the goals that they satisfy and the state of the world for which they are valid.

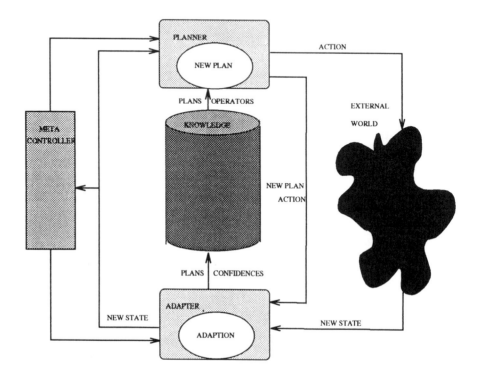

Figure 4.4: Overview of Planner architecture.

- **Generalizer**. The Generalizer modifies the plan knowledge in the knowledgebase by simpifying it. Related plans are merged to form an abstract plan with a wider scope of applicability. This helps to reduce the number of stored plans.

The architecture is designed to support knowledge storage and access with the ability to impose context sensitive planning through a Meta-Controller which can modify the planning and learning mix. Learning can then be given more computational time during reasonably uneventful periods, while planning can be given priority during busy periods.

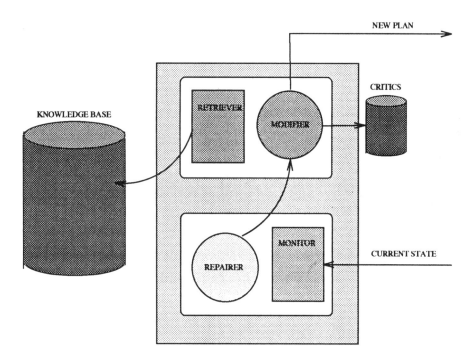

Figure 4.5: Planner architecture.

4.8   Conclusions

The main objective in this work was to integrate recent developing technologies in control and knowledge based systems in order to address the long standing difficulty with the automation of management systems. Specifically, neurofuzzy approaches to control and case based planning. Neurofuzzy networks can be expressed in a manner suitable for knowledge based systems development and particularly case based planning. A neurofuzzy approach developed by Brown [9] has been adopted. This approach uses B-spline continuous distributions which help to impose a partition of unity across individual dimensions, and the use of product(norm) and normalized addition(conorm) for representing the AND and OR, respectively, to ensure a smooth output

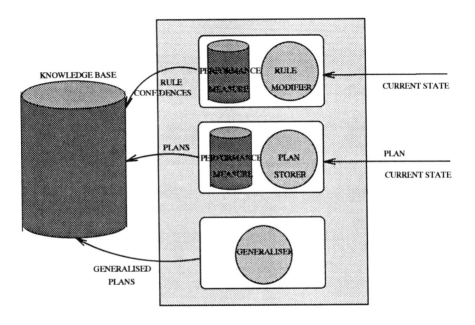

Figure 4.6: Adapter architecture.

while utilizing all input information. The use of normalized 'degrees of confidence' allow the fuzzy rule base to behave in a neural net like fashion with its associated advantages.

Developing models that correctly reflect the real world is only possible for tightly constrained domains. Given this, a situated agent like an IAV can't rely too heavily on its internal model of the world, and should alternatively rely more on experience. Case based planning adopts this approach where experience based knowledge is stored, retrieved and modified to suit different situations. Causal models are represented by fuzzy subnetworks which map some set of input relations into one or more output relations. The adaption of these causal models is through the relative emphasis or de-emphasis of fuzzy rules in individual fuzzy subnetworks through the modification of their associated 'degrees of confidence'. This modification is driven by a rule-based performance measuring surface, which relates the difference between expected results of performing an action and the actual results, to changes in rule confidences. Strategic plans describe input/output relations for sequences of actions. They represent acquired skills which are context sensitive and specific to a set of goals. They are important in that they reduce search, and therefore computational load, by chunking actions together. Abstract plans are plans which have been abstracted from acquired strategies. The abstract plan maps inputs to sequences of actions, but ignores inputs which are deemed insignificant. As such, abstract plans are enormously important to the process of outlining future activities, and thus to the process of interleaving planning and execution.

The maturing knowledge based planning field has adopted new architectures for planning which are more sympathetic to situated activity. With the adoption of a neurofuzzy technique, it is believed that some headway may be made into the problematic development of management systems, and specifically Mission Management systems for MIAVs.

References

[1] Albus J.S., McCain H.G. and Lumia R. (1989) NASA/NBS Standard Reference Model for Telerobot Control System Architecture (NASREM). *NIST-TN-1235*, National Institute of Standards and Technology, USA.

[2] Albus J., Quintero R., Lumia R., Herman M., Kilmer R. and Goodwin K. (1990) Concept for a Reference Model Architecture for Real-Time Intelligent Control Systems (ARTICS). *NIST-TN-1277*, April.

[3] Albus J.S. (1993) A Reference Model Architecture for Intelligent Systems Design. In Antsaklis P.J. and Passino K.M. (eds) *An Introduction to Intelligent and Autonomous Control*. Kluwer Academic, Chapter 2, 27–56. ISBN 0-7923-9267-1.

[4] Baldwin J.F. (1979) Fuzzy Logic and Fuzzy Reasoning. *Fuzzy Logic and Reasoning.*

[5] Baldwin J.F. (1985) Fuzzy Sets and Expert Systems. *Information Sciences.* (6).

[6] Baldwin J.F. (1986) Support Logic Programming. *Fuzzy Sets Theory and Applications*, Jones A. (ed).

[7] Berenji H.R. (1992) Fuzzy Logic Controllers. In *An Introduction to Fuzzy Logic Applications and Intelligent Systems*, Yager R.R, Zadeh L.A (eds), Kluwer Academic Publisher, Boston, MA, Ch.4.

[8] Bossley K.M., Mills D.J., Brown M. and Harris C.J., (1994) Neurofuzzy high dimensional approximation. In *Neural Networks*, J.G.Taylor (ed),

297–332. Alfred Waller Ltd, Henley-on-Thames in association with UNI-COM.

[9] Brown M. and Harris C. (1994) *Neurofuzzy Adaptive Modelling and Control.* Prentice Hall, Hemel Hempstead.

[10] Brooks R.A. (1986) A Robust Layered Control System for a Mobile Robot. *IEEE J. of Robotics and Automation* (RA-2) 1.

[11] Corfield S.J., Fraser R.J.C. and Harris C.J. (1991) Architectures for Real-Time Control of Autonomous Vehicles. *IEE Computing and Control Engineering Journal* (2) 6: 254–262.

[12] Daniel L. (1983) Planning and Operations Research. *Artificial Intelligence: Tools, Techniques and Applications.* Harper and Row, New York.

[13] Dubois D. and Prade H. (1980) *Fuzzy Sets and Systems: Theory and Applications.* Academic Press, London.

[14] Fikes R.E. and Nilsson N.J. (1971) STRIPS: A New Approach to the Application of Theorem Proving to Problem Solving. *Artificial Intelligence* 2: 189-208.

[15] Fikes R.E., Hart P.E. and Nilsson, N.J. (1972) Learning and Executing Generalised Robot Plans. *Artificial Intelligence* (3).

[16] Fikes R.E., Hart P.E. and Nilsson, N.J. (1972) Some New Directions in Robot Problem Solving. *Machine Intelligence* (7) Edinburgh University Press.

[17] Fraser R.J.C., Harris C.J., Mathias L.W. and Rayner, N.J.W. (1991) Implementing Task-Level Mission Management for Intelligent Autonomous Vehicles. *Engineering Applications of Artificial Intelligence* (4) 4: 257–268.

[18] Hammond K. (1988) Case Based Planning: Viewing Planning as a Memory Task. *Proceedings Case Based Reasoning Workshop.* Morgan Kaufmann, May.

[19] Hammond K. (1989) *Case Based Planning: Viewing Planning as a Memory Task.* Academic Press, Boston.

[20] Harris C.J., Moore, C.G. and Brown M. (1993) *Intelligent Control: Aspects of Fuzzy Logic and Neural Nets.* World Scientific, London.

[21] Harris C.J. (1991) Advances in Intelligent Autonomous Vehicles. *Editorial Special Issue. En. Applic. AI Journal* (4) 4: 251–255.

[22] Harris C.J. and Fraser R.J.C. (1992) Command and Control for Intelligent Autonomous Vehicles: an approach emphasising interoperability. *12th IFAC Symposium on Aerospace Control.* Ottobrunn, Germany, 61–70.

[23] Harris C.J. and Fraser R.J.C. (1993) Command and Control Infrastructures: The Need for Open Systems Solutions. *4th International Conference on Advances in Communication and Control* (COMCON 4), June 14th - 18th, Rhodes, Greece.

[24] Kavli T. (1992) *Learning Principles in Dynamic Control*, PhD thesis, University of Oslo, Norway.

[25] Kavli T. (1993) ASMOD: an algorithm for Adaptive Spline Modelling of Observation Data. *International Journal of Control* (58) 4: 947–968.

[26] Marks M., Hammond K.J. and Converse T. (1988) Planning in an Open World: A Pluralistic Approach. *Proceedings Case Based Reasoning Workshop.* DARPA May.

[27] Meyer B. (1988) *Object Oriented Software Construction.* Prentice Hall.

[28] Meyer B. (1989) The New Culture of Software Development: Reflections on the Practise of Object Oriented Design. *Proceedings of Technology of Object Oriented Languages and Systems (TOOLS-89)*, Paris, France, November 13-15, 13–23.

[29] Pedrycz W. (1993) *Fuzzy Control and Fuzzy Systems.* 2nd ed. Research Studies Press, Taunton, John Wiley and Sons.

[30] Procyk T.J. and Mamdani E.H. (1979) A Linguistic Self Organising Controller. *Automatica* (15) 15–30.

[31] Ram A. and Santamaria J.C (1993) Multistrategy Learning in Reactive Control Systems for Autonomous Robotic Navigation. *Informatica* (17) 4: 347–369.

[32] Sacerdoti E.D. (1973) Planning in a Hierarchy of Abstraction Spaces. *IJCAI-73*, Palo Alto, CA.

[33] Sacerdoti E.D. (1975) The Non-linear Nature of Plans. *IJCAI-77*, Tiblisi, USSR.

[34] Sacerdoti E.D. (1977) *A Structure for Plans and Behaviour.* Elsevier.

[35] SAGEM (1993) PANORAMA: Final Edited Public Report. *PANORAMA Project Document*, Ref: EPPR/SAG931130, October.

[36] Segre A.M. (1988) *Machine Learning of Robot Assembly Plans.* Kluwer Academic, Norwell.

[37] Soley R.M. (1993) OMG: Creating Consensus in Object Technology. *Object Management Group Presentation.*

[38] Sun C. (1994) Rule-base structure identification in a adaptive network based inference system. *IEEE Transactions on Fuzzy Systems* (2) 1: 64–73.

[39] Taylor J.G. (1995) (ed) (1995) *Neural Networks*. Alfred Waller.

[40] Wang L.X. (1994) *Adaptive Fuzzy Systems and Control: Design and Stability Analysis*, Prentice Hall, Englewood Cliffs, NJ.

[41] Wilkins D.E. (1988) *Practical Planning*. Morgan Kaufmann, San Mateo, California.

[42] Zadeh L.A. (1978) Fuzzy Sets as a Basis for a Theory of Possibility. *Int.J.Fuzzy Sets Syst* (1) 1: 3–28.

[43] Zadeh L.A. (1964) Fuzzy Sets. *Memo.ERL, No. 64-44*. Univ. of California, Berkeley.

5

A Fuzzy Data Browser in Fril

J.F. Baldwin, T.P. Martin

5.1 Introduction

The emerging technologies of soft computing promise to solve many real-world
problems which are difficult to tackle using conventional approaches. Until
now the most prominent application area has been fuzzy control, but there
is scope for a much wider utilization of fuzzy techniques in knowledge bases.
The Fuzzy Data Browser is a Fril program designed to assist a human user
in extracting relationships from large bodies of data. There are many reasons
why one would wish to extract rules from data — a rule is easier for a human
to understand than a large table of data, and the derivation of rules can
confirm or enhance a user's understanding of the deeper structure underlying
the raw data. Rules are more compact, and can be used as a form of data
compression.

In many cases, data may not be complete — for example, weather records
could include daily temperature extremes, rainfall, cloud cover, etc. If a new
machine became available for monitoring atmospheric pollution, these measurements could be added to the database, but the earlier records would not
contain this information. If there is a relation between atmospheric pollution
and some of the other recorded quantities, rules modelling this relation could
be used to make an intelligent guess as to the pollution level for each record
where it was not measured.

Finally, if rules can be derived which summarize the underlying data, it is
possible to use the rules to highlight anomalous values — for instance, a bank
could maintain a rule summarizing the spending pattern of a customer; each
transaction could be compared to a predicted pattern, and any transaction

which was 'unexpected' could be referred to a supervisor for scrutiny.

The Fuzzy Data Browser will extract rules from data, and can be used with various levels of human input. The system can run entirely autonomously, or the user can inject expertise either in:

- suggesting attributes which should be the target for prediction;

- suggesting attributes which should be used to make the prediction;

- suggesting compound features (i.e. transformations of the original attributes) which could form a better basis for the prediction;

- examining and editing the rules generated by the system.

Fril [11, 12, 13] combines uncertainty and logic programming in an AI language with powerful and flexible features for handling uncertainty.

Most knowledge-based systems need to deal with the problem of uncertainty in developing real-world applications, and consequently there is a need for software tools able to address this problem. Fril is a commercially available AI language which handles uncertainty as an intrinsic part of the language. It is based on logic programming, extended by the concept of a mass assignment as the fundamental means of representing uncertainty. Mass assignments unify probability and possibility theory and form a coherent mathematical framework for dealing with uncertainty in knowledge-based systems. Mass assignments and the theoretical foundations of Fril have been explained in several other papers [4, 5], and are not covered in depth here. Instead, we concentrate on practical aspects of Fril, illustrated by simple examples.

Logic programs consist of facts and rules. Programs are executed by posing a query, which is answered by determining whether that query (or some instance of it) is a logical consequence of the clauses in the knowledge base. Execution is a search over an *and-or* tree. Fril extends logic programming (Prolog) by allowing fuzzy sets as fundamental data objects, and associating a support pair with each fact or rule. A support pair is an interval containing a point-value probability. The probability can take any value in the interval, avoiding the need to specify artificially precise point values. Execution of a support logic program is also a search over the *and-or* tree, with additional computation required to calculate supports associated with each solution.

Fril is implemented as a compiler for the abstract Fril machine [8], which is a modification of the Warren Abstract Machine [17] used in most Prolog compilers. This yields a highly efficient support logic programming system. Clauses (possibly involving supports) are compiled into sequences of abstract machine instructions, which can be executed by a software emulator or by a hardware implementation of the abstract machine.

Fril has been used in a wide range of industrial and academic projects which have needed to model uncertainty. This chapter gives an overview of the language, concentrating on the features used to handle uncertainty, and outlines a practical application of Fril in aiding the design of fuzzy data models.

5.2 Mass Assignments

A brief introduction to mass assignments is necessary to appreciate the methods used to form fuzzy sets from data. A mass assignment m is defined over the set X by the function

$$m : P(X) \rightarrow [0, 1]$$

where

$$\sum_{A \in P(X)} m(A) = 1 \text{ and } m(\emptyset) \geq 0.$$

Any mass assignment represents a family of distributions $\{FD(x_1), \ldots, FD(x_n)\}$ over the universe of discourse $X = \{x_1, \ldots, x_n\}$ where

$$m(\{x_i\}) \leq FD(x_i) \leq \sum_{A = \{x_i\} \cup Y} m(A)$$

with the constraint

$$\sum_i FD(x_i) = 1 - m(\emptyset) \quad ; \quad \forall x_i \in X$$

if there is no mass on the null set then this family will be exactly equivalent to a family of probability distributions over X.

We distinguish a single distribution from this family, the *least prejudiced distribution*, obtained by distributing the mass associated with any subset A equally between its elements. For example, we may consider a case where objects are classified on a production line as either oval, circular or rectangular. For any given batch of 100 components, it is known that 30 are circular, 40 are rectangular or circular, and the remainder are unknown. We can represent this by a mass assignment with the domain $X=\{$oval, circular, rectangular$\}$ as

$$\text{batch} = \{\text{circular}\} : 0.3, \{\text{rectangular, circular}\} : 0.4,$$

$$\{\text{rectangular, oval, circular}\} : 0.3.$$

Although mass assignments can represent probabilities they have the added flexibility of being able to represent uncertain probabilities. This mass assignment represents the following family of probability distributions:

$$0.3 \leq \text{Pr(circular)} \leq 1$$
$$0 \leq \text{Pr(rectangular)} \leq 0.7$$
$$0 \leq \text{Pr(oval)} \leq 0.2$$

such that

$$\text{Pr(circular)} + \text{Pr(rectangular)} + \text{Pr(oval)} = 1.$$

The least prejudiced distribution is obtained by equally dividing the mass on the non-singleton subsets among their elements; thus we obtain

$$\text{Pr(circular)} = 0.3 + 0.4/2 + 0.3/3 = 0.6$$
$$\text{Pr(rectangular)} = 0.4/2 + 0.3/3 = 0.3$$
$$\text{Pr(oval)} = 0.3/3 = 0.1.$$

The transformation to least prejudiced distribution is reversible; hence given a least prejudiced distribution, we can find a corresponding mass assignment.

Mass assignments are related to fuzzy sets via the voting model [2] as follows:

Suppose that

$$V \text{ is } f$$

where f is a fuzzy set defined on the discrete space $X = \{x_1, x_2, \ldots, x_n\}$, namely

$$f = \sum_{i=1}^{n} x_i/\chi_i$$

then the fuzzy set f induces a possibility distribution over χ for the variable V, namely

$$\Pi(x_i) = \chi_i$$

Suppose f is a normalized fuzzy set whose elements are ordered such that

$$\chi_1 = 1, \quad \chi_i \geq \chi_j \quad \text{if} \quad i < j$$

then

$$\Pi(\{x_i, \ldots, x_n\}) = \chi_i$$

so with the assumption that $\text{Pr}(A) \leq \Pi(A)$ for any $A \in P(X)$ we can find that the mass assignment corresponding to the fuzzy set f is

$$m_f = \{\{x_1, \ldots, x_i\} : \chi_i - \chi_{i+1}\} \text{ with } \chi_{n+1} = 0.$$

This can be extended to non-normalized fuzzy sets so that the mass assignment corresponding to the fuzzy set f if f is non-normalized is

$$m_f = \{\{x_1, \ldots, x_n\} : \chi_i - \chi_{i+1}, \{\emptyset\} : 1 - \chi_1\} \text{ with } \chi_{n+1} = 0,$$

such that a non-zero mass is assigned to the null set, in this case the mass assignment is said to be incomplete. The extension to fuzzy sets over multiple domains is straightforward.

The relationship between probability and possibilities has been investigated by others including [21, 19, 15]. Taking the fuzzy set **low-numbers** defined on the universe $\{1, 2, 3, 4, 5, 6\}$

$$\textbf{low-numbers} = 1/1 + 2/1 + 3/0.5 + 4/0.2 + 5/0 + 6/0$$

the mass assignment of the fuzzy set **low-numbers** is

$$m_{\text{low-numbers}} = \{1, 2\} : 0.5, \quad \{1, 2, 3\} : 0.3, \quad \{1, 2, 3, 4\} : 0.2.$$

We see that the sets are nested which will always be the case when converting fuzzy sets to mass assignments. Thus, there is a straightforward transformation from frequency distributions to mass assignments and then to fuzzy sets. The examples have illustrated the discrete case; the continuous case is similar.

5.3   Key Features of Fril

We illustrate Fril using a simple database. A relational database imposes a crisp model of the world, *i.e.* all categories must be precise and information must be known with complete certainty. In practice this leads to arbitrariness. For example suppose we have information such as:

- Mary is either 28 or 29;

- Mary is in her late twenties;

- Mary is thought to be 29 (not certain);

- Mary is thought to be in her late twenties (again, not known for certain).

This information is difficult to model in a relational database as it is imprecise, uncertain, or both. The information stored in the database is forced into precise and certain categories, even when this is not justified by the information known about the real world. A closely related problem is that 'soft' queries are not permitted, *e.g.* the database cannot answer questions such as:

- Which small departments are *highly efficient*;

- In which departments does the number of staff *considerably exceed* agreed levels;

- How many employees *nearing retirement age* earn *large salaries*

without arbitrary definitions of the terms in italics. For example, one answer to the last question could be obtained by defining nearing retirement age as greater than 60 and a large salary as more than £50,000; defining the cut-off points at 55 and £60,000 could lead to a very different answer. From the logical point of view, it is necessary to define precise categories but this can lead to a discord between the real world and the model in the database, as the terms used to describe the world are naturally imprecise.

A good solution to this problem is to use fuzzy sets to model the imprecise terms. We distinguish two uses of fuzzy sets in this context:

(i) when the item in question is single-valued but is not known precisely. For example, the speed of a car might be described as *around 75mph* This gives a possibility distribution of speeds; in principle it is possible to measure the speed accurately and obtain a single value. Using a discrete fuzzy set for simplicity, the statement

$$\text{speed of car-1 is } \{70: 0.5, 75:1, 80:0.2\}$$

represents a *disjunction* of statements with memberships in the set of true statements:

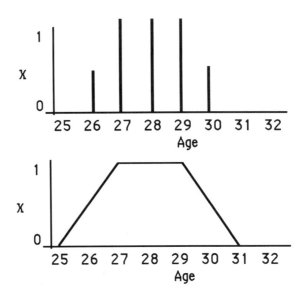

Figure 5.1: Discrete and continuous fuzzy sets representing the age late twenties.

$$
\begin{array}{rl}
      & \text{speed of car is 70 (membership 0.5),} \\
\text{or} & \text{speed of car is 75 (membership 1.0),} \\
\text{or} & \text{speed of car is 80 (membership 0.2),}
\end{array}
$$

(ii) when the item in question is set-valued, but the boundaries of the set are not known precisely. For example, the safe speed on a road could be *around 75 mph*. This represents a *conjunction* of statements with associated memberships:

$$
\begin{array}{rl}
       & \text{safe speed on Road1 is 70 (membership 0.5),} \\
\text{and} & \text{safe speed on Road1 is 75 (membership 1.0),} \\
\text{and} & \text{safe speed on Road1 is 80 (membership 0.2).}
\end{array}
$$

Case (i) is modelled in Fril by a fuzzy set as data value; case (ii) is modelled by supported clauses (see also [15, 20]).

5.3.1 Uncertainty in Data Values

Uncertain data values such as Mary's age (above) can be modelled using a fuzzy set in Fril. We might adopt either of the definitions (see Figure 5.1):

(late-twenties {26:0.5 27:1 28:1 29:1 30:0.5})
(late-twenties [25:0 27:1 29:1 31:0])

depending on whether the domain of age is the discrete set of positive integers
or the real line. The notation *element : membership* is used for element-
membership pairs in fuzzy sets. Square brackets [ ] indicate that the set is
on a continuous domain, with linear interpolation between adjacent points;
braces { } indicate that the set is on a discrete domain. Fril allows either
named or unnamed fuzzy sets. Thus we could define the named set *late-
twenties*, and use the fact

((age Mary late-twenties)).

Alternatively, we could simply use the fact

((age Mary [25:0 27:1 29:1 31:0])).

If the domain is not numerical, Fril fuzzy sets can still be used, *e.g.* a fuzzy
set of small cars:

(small-cars mini:1 metro:0.9 escort:0.3).

Continuous fuzzy sets can be used in arithmetic expressions in Fril, *e.g.* if
information on the ages of three employees is represented by the facts:

((age John 33))
((age Mary [25:0 27:1 29:1 31:0]))
((age Bill 42))

and we wish to find the average age, we could find a list of ages (33 [25:0
27:1 29:1 31:0] 42), sum the ages and divide by the number of elements in the
list. The total is [100:0 102:1 104:1 106:0], which gives [33.33:0 34:1 34.67:1
35.33:0] on division by 3. Thus we can combine precisely known information
with approximate information and derive a solution which is imprecise but
is nevertheless useful, *e.g.* if we needed to know whether the average age
was below 40, we could answer the question with complete certainty. The
built-in arithmetic predicates of Fril allow continuous fuzzy sets to be used as
arguments.

5.3.2   Semantic Unification

Of course, we need to model fuzzy queries as well as fuzzy data values — for
example, consider the query *find members of staff whose age is near 60*. If
data is precisely known, this involves matching a crisp value with a fuzzy set;
however, if the data is fuzzy, it involves matching two fuzzy sets, *e.g.* given that
Fred is *about 57*, what is the support for him being *near 60* (see Figure 5.2).
This process of matching fuzzy sets is known as *semantic unification* and is a
fundamental part of Fril. A support pair is automatically calculated for the
match, and incorporated into the overall calculation of support for the query,
using (by default) the probabilistic semantic unification model. Alternative
models within the system provide possibilistic matching, and a refinement of
the probabilistic matching which gives a point value instead of an interval
using an intelligent algorithm to fill in missing data. Further details of all
methods are given in [6, 7].

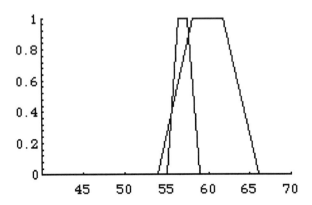

Figure 5.2: Continuous fuzzy sets representing the ages *about 57* and *near 60*.

5.3.3   Uncertainty in Facts

In logic programming, a fact represents a relation between objects; the predicate names the relation and the arguments name the objects satisfying the relation. Fril facts generalise relations by including a support pair for each tuple. Thus if we have domains *road* and *speed* we can define a relation

$$\text{safe-speed} \subseteq f \text{ road } x \text{ speed}$$

where for any tuple there is no uncertainty in the attribute values, but there may be uncertainty as to how well a pair of values satisfies the relation.

For example, travelling on a minor road at 90 mph is definitely not safe, but travelling on a motorway at 65 mph is reasonably safe. Other tuples have different degrees of membership. There is no uncertainty in the attribute values; the uncertainty is the degree to which a tuple satisfies the relation. Fril models uncertainty in a relation by support pairs. These are more general than memberships in a fuzzy relation, although by restricting supports to point values and choosing an appropriate calculus, Fril can easily model fuzzy relations. Discussion of the semantics of support pairs can be found in [13]. Discrete fuzzy relations can be entered directly into Fril; continuous fuzzy relations can also be modelled but are best written using rules (section 5.3.4), *e.g.*

((safe-speed-on motorway is SPD)
    (match [60:0 70:1 80:1 90:0] SPD)) : ((1 1)(0 0))

where we use semantic unification to determine the support for SPD matching the safe range defined by the fuzzy set [60:0 70:1 80:1 90:0]. The support pair on the rule is an equivalence, giving the same support to the head of the rule as to the body.

Fril can combine uncertainty in attributes and in the relation between attributes. For example, 'Mary is strongly believed to be in her late twenties' could be represented as

$$((age\ Mary\ [25:0\ 27:1\ 29:1\ 31:0])) : (0.8\ 1)$$

where the support pair (0.8 1) represents the qualifier *strongly believed*.

5.3.4 Rules

Fril can model uncertainty in rules as well as in data items and facts. There are a number of ways in which uncertainty can be processed in rules; here, we focus on the default calculus and the evidential logic method. It is possible to customize the calculus for dealing with uncertainty, so that (for example) a fuzzy logic style max-min calculus could be defined for a particular rule. To illustrate the basic rule in Fril, suppose we have a set of facts giving details of various companies:

$$((turnover-of\ `acme\ plc'\ in\ 1992\ is\ 200000))$$
$$... \ etc.$$
$$((profit-of\ `acme\ plc'\ in\ 1992\ is\ 12\ \%))$$
$$... \ etc.$$

We could define a rule stating that a company performs well in a particular year if it makes a good profit and has a high turnover:

company X performed well in YEAR
     IF   turnover-of X in YEAR is *high-turnover*
  AND profit-of X in YEAR is $\approx$ 10-20 %
This is expressed in Fril as
$$((company\ X\ performed\ well\ in\ YEAR)$$
$$(turnover-of\ X\ in\ YEAR\ is\ high\text{-}turnover)$$
$$(profit-of\ X\ in\ YEAR\ is\ about\ 10\text{-}20\ \%)) : (0.8\ 1)$$

where the italicised terms are fuzzy sets. The support pair (0.8 1) reflects the heuristic nature of the rule — if the conditions are true, the head will be true with a probability in the interval [0.8 1]. A rule support generally consists of two pairs — one specifying an interval for the head being true when the body is true, the other specifying the interval for the head being true when the body is false. In general, the support for the body is between true and false, and both pairs are used to compute an overall support for the head. If the second support pair is omitted it defaults to uncertainty, *i.e.* (0 1). The body support is calculated as the product of the support pairs for each goal in the body.

5.3.5   Evidential Logic Rule

In the rule above, Fril produces an uncertain result if one of the conditions is found to be false or nearly false. We may wish to be more tolerant of failures in a rule's conditions, *e.g.* we could say that a company has a good track record if it has performed well in recent years. However, a failure to perform well in a single year should not exclude the company completely — let us say a company should satisfy the rule if it has performed well in *most* of the last four years. The evidential logic rule in Fril allows this tolerant behaviour:

((company X has good track record)
  (evlog *Most*
          ((company X performed well in 1994) 0.4
          (company X performed well in 1993) 0.3
          (company X performed well in 1992) 0.2
          (company X performed well in 1991) 0.1)))) : ((1 1) (0 0))

where the built-in predicate evlog indicates that the evidential logic calculus is used for this rule. This allows us to weight the importances of the conditions, and also allows the rule to tolerate the failure of a condition (particularly one of the less important conditions) without completely degrading the support for the conclusion. The body support is calculated as the weighted sum of support pairs for each goal, and the weighted sum is scaled according to the fuzzy set filter — *Most* in the example above.

The evidential logic rule is particularly appropriate for classification problems where most of a set of features are required to be present, but the lack of a single feature should degrade the support for that classification slightly, rather than preventing the conclusion from being drawn. The evidential logic rule has been successfully applied in a number of areas, including identification of underwater sounds, prediction of aircraft performance data, and the selection of models for safety assessment.

5.3.6   Additional Features

In addition to the features described above for dealing with uncertainty, Fril also contains a complete Prolog system, with a list-based syntax. This includes many program development tools, including

- full support for tracing/debugging programs;

- creation of self-contained code modules (in which further optimizations are possible);

- the ability to create stand-alone applications (in which the user sees only the interface provided by the application, and is unaware of the underlying Fril system);

- the ability to link with code in other languages. This can be on a tightly coupled basis, where Fril calls functions defined in another language, or where Fril is embedded within a larger framework and is called as a module within a large package. Alternatively, the linkage can be looser, with Fril and another application running as separate communicating processes. A good example is the interface to Mathematica, where Fril and Mathematica run separately (on the same or different machines) but exchange data and results to solve a problem [9].

5.4 Fuzzy Data Browser

In many large databases, information may be incomplete or uncertain. For example, consider a database recording the results of scientific experiments assembled from published papers. The experiments may use different approaches, so that results may not be directly comparable across the whole set of cases; we might like to predict the result of an existing experiment if conditions had changed slightly, or a slightly different method had been used. New equipment may enable more data to be collected; we might like to know what the extra values would have been if the new equipment had been available when the original experiments were carried out.

The Fril data browser can be used to form rules which predict the unknown values in an experiment from the known values in that experiment, plus the values in other similar experiments. The database is partitioned into fuzzy subsets containing similar values of the variable under consideration; each rule then uses fuzzy sets to summarize values of other variables in that partition.

Let the database be a relation on $D_1 \times D_2 \times \cdots \times D_n$

$$R \subseteq D_1 \times D_2 \times \cdots \times D_n$$

$$R = \{t_i | i = 1, \ldots, m\}$$

where $t_i = (a_{i_1}, a_{i_2}, \ldots, a_{i_n})$ such that $a_{i_1} \in D_1 \vee a_{i_1} \subseteq_f D_1$.

This can be extended to fuzzy subsets of $D_1 \times D_2 \times \cdots \times D_n$ includes cases where attribute values a_{i_j} are set-valued (including fuzzy subsets), as described in the previous section. Let us assume that we wish to predict the value of some attribute A_j. We must first form a fuzzy partition of the domain D_j

$$P_j = \{H_{1_j}, H_{2_j}, H_{m_j}\} \quad \text{where} \quad H_{i_j} \subseteq_f D_j.$$

This fuzzy partition can be used to group the tuples:

$$G1 = \{(a_{i_1}, a_{i_2}, \ldots, a_{i_n})/\mu_{i_1} | a_{i_j} \in H_{1_j} \quad \text{with} \quad \text{membership} \quad \mu_{i_1}\}$$

etc. The case where a_{i_j} is a fuzzy set is easily dealt with.

This gives a fuzzy partition of the database. Each fuzzy subset in the partition can be converted to a least prejudiced distribution, and a corresponding frequency distribution on each attribute can be extracted. These are converted back to fuzzy sets, which give an approximate value for each

Table 5.1: Example data for the fuzzy circle problem.

Number	x value	y value	Classification
1	0.3	0.5	valid
2	0.7	1.1	invalid
3	about-0.1	0.9	valid
4	about-0.4	about-0.9	invalid

attribute summarizing the cases in that element of the partition. The browser then yields a set of rules of the form value of A_j is H_{k_j} IF value of A_1 is F_{k_1} AND value of A_2 is F_{k_2} AND ... where F_{k_i} are the fuzzy sets found from the data. It is possible to generate ordinary support logic rules, or to use evidential logic rules, in which case the importance of each feature can be found using semantic discrimination analysis. In this case, it is also possible to discard unimportant features, since they are found to have very low importance, *i.e.* they give similar fuzzy sets across all fuzzy classes in the partitions, and are not good at discriminating.

The value of attribute a_{i_j} in the tuple $(a_{i_1}, a_{i_2}, \ldots, a_{i_n})$ can be predicted using these rules. By executing the rules, a support pair S_i is found for each fuzzy class H_{i_j}. If a classification is required (*e.g.* the value is small, medium, or high) then this is directly available by comparing the support pairs. On the other hand, a point value may be required. This can be extracted by converting the support pairs and fuzzy classes into an expected fuzzy set and then taking the expected value from the corresponding least prejudiced distribution.

5.4.1   Example

We illustrate the use of the Fuzzy Data Browser in a simple, artificial problem, illustrated in Figure 5.3. A random selection of points are classified as legal or illegal, according to whether they fall inside or outside a quarter-circle of radius 1. Some points near the boundary have been fuzzified, *i.e.* either the x value or the y-value, or both, are known fuzzily. This simulates the situation in real-world situations, where borderline data is subject to uncertainty. The database consists of four columns, (see Table 5.1), and there are 200 tuples. The goal is to devise rules which can predict the classification (valid or invalid). Note that in this case the classification is crisp, and only the data is fuzzy — the case of a fuzzy classification is just as simple to handle in the Fuzzy Data Browser.

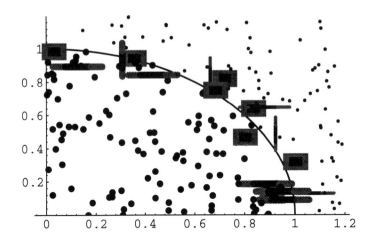

Figure 5.3: Artificially generated data for a fuzzy circle. Large dots correspond to valid points, small dots to invalid points. Some points are fuzzy in their x and/or y values; these are shown as a black area (membership 1) and grey areas (membership decreasing to zero).

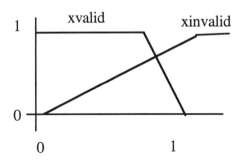

Figure 5.4: Fuzzy sets for **xvalid** and **xinvalid** generated by the data browser.

5.4.2   Use of the Fuzzy Data Browser

Using the x and y features given in the database, a simple fuzzy rule for a valid point is:

((classification of point N is valid) /* IF */
      (xvalue of point N is **xvalid**)
      (yvalue of point N is **yvalid**))

where **xvalid** and **yvalid** are fuzzy sets derived from the data. Similar rules are obtained for the classification invalid, and testing the predicted against actual values gives a success rate of 89%. Essentially, the two fuzzy sets on x and y define a fuzzy square, and a number of points are mis-classified because the coupling between x and y values is not modelled by treating the x and y features independently. The success rate of 89% can be improved by defining a compound feature, which is a transformation of the simple features present in the database. There are obviously very many possibilities for combining x and y, and in the absence of other knowledge, we could use a genetic algorithm to generate and evaluate new features. However, given our background knowledge, we can create the new feature *xysqyare*, defined as $x^2 + y^2$, which yields rules of the form

```
((classification of point N is valid)     /* IF */
       (xysquare of point N is xyvalid))
((classification of point N is invalid)   /* IF */
       (xysquare of point N is xyinvalid))
```

This gives a success rate of 100%.

Note that to model this problem and create a new feature which is a simple function of existing attributes, we need a system which can handle fuzzy data values, and also perform fuzzy arithmetic on those values. Fril is therefore ideal.

5.5   A Real-World Problem — Database of Experimental Measurements

5.5.1   The Problem

The measurement of aquifer dispersivities is a significant component in modelling the flow of water in rock formations. Predicting the movement and spread of contamination in water supplies is an important application where aquifer dispersivities must be determined accurately. Unfortunately, experimental measurements show a scale dependence which is not predicted by theory; thus values determined experimentally in laboratories are not generally useful in predicting values to be used in large scale calculati ons. Some experiments have been performed in the field (*i.e.* measurements are taken on real aquifers, rather than in the laboratory) and these have been examined closely for use in predicting values to be used in calculations. For example, [16] examined a number of field experiments and tabulated over 100 results. This table forms our database, which exhibits:

- discrete and continuous data;

- incompleteness (not all experiments take the same set of measurements, so there are gaps in the data);

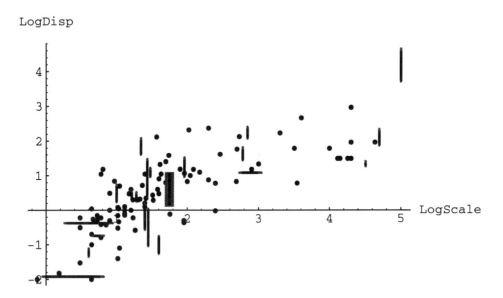

Figure 5.5: Log Dispersivity (y axis) plotted against Log Experiment Scale (x axis) for the Dispersivity database. Points which are fuzzy in their x and/or y values are shown as a black area (membership 1) and grey areas (membership decreasing to zero). We note that one very uncertain point has been omitted (see Figure 5.7(a)).

- uncertainty (many values are quoted as ranges of possible values);

- unreliability (the authors classify each experiment as high, medium, or low reliability).

All of these features are easy to model in Fril. Where ranges appear in the data, these have been modelled by possibility distributions. There are 18 attributes in the database, and 116 tuples. The database contains the following attributes:

Site and Experiment Number unique identifier for each row in the database
(57 sites).

Aquifer Thickness	[0 – 1000] m
Experiment Scale	[0 – 100 000] m
Longitudinal Dispersivity	[0 – 45000]
Transverse Dispersivity	[0 – 1500]
Vertical Dispersivity	[0 – 1]
Material	Sand, Gravel, Alluvial Deposits,
	...(26 rock types in total))

Effective Porosity	[0–100]%
Hydraulic Conductivity	$[10^{-8} - 10^{-1}]$
Hydraulic Transmissivity	$[10^{-8} - 20]$
Velocity	$[0 - 220]$
Flow Configuration	Ambient, Radial Converging, Radial Diverging
	...(8 in total))
Monitoring	two-dimensional or three-dimensional
Tracer	Br-, Tritium, fluorescein,...(29 in total)
Input Method	pulse, contamination, step, or environmental
Data Interpretation	2-D Numerical, 1-D Uniform Flow Solution,
	...(15 in total)
Reliability	low, reasonable, or high

Because of the large range of the Experiment Scale and Longitudinal Dispersivity attributes, we work with the log of the value rather than the value itself. This is an example of defining a derived relation in Fril, where we work with a value calculated from the data, rather than a value that is explicitly present. Although taking the log of a value is a relatively easy operation, it should be emphasised that the browser enables complete freedom in defining new attributes and relations on the existing data.

Much of the data is incomplete, since investigators use different experimental set-ups and methods of analysing data. There is also uncertainty in values, due to the difficulty in measuring or estimating data. Finally, we note that the database is defined on both discrete and continuous domains.

5.5.2   Use of the Fuzzy Data Browser

To illustrate the Fuzzy Data Browser, we predict one attribute (longitudinal dispersivity) from other attributes which can be chosen either by 'expert intuition' or by examination of the data. As a rule of thumb, it is generally reckoned by experts that the scale of the experiment and the longitudinal dispersivity are approximately linearly related; however, because both quantities vary over several orders of magnitude in the database, this relation is difficult to see graphically. Having defined the database to the data browser, our first move is to define two derived features, Log(Dispersivity) and Log(Experiment Scale). These can be displayed graphically using a link with Mathematica (see Figure 5.5). It is apparent that there is a rough linear trend, and this can be used to form rules which predict LogDisp from LogScale.

The relationship between other attributes and the Log(Dispersivity) can be examined using the data browser to construct predictive rules. We consider two ways of determining the most important features. In the first approach, we extract important features by analysis of the weights given to each feature by the browser. Initially, five evidential logic rules were formed, using all available attributes. The LogDispersivity domain is split into five classes LogDispClass1-5, as in Figure 5.6. Each rule then has the form

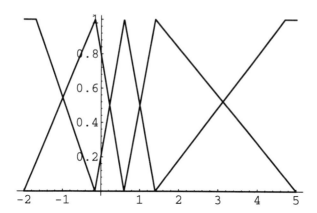

Figure 5.6: Fuzzy sets on the Log Dispersion domain. The sets were chosen automatically by the system, to ensure a roughly equal number of cases in each class.

((Predicted value for LogDisp in case (Site ExperimentNumber) is
                    LogDispClass_i)
    (evlog
        ((value of LogScale in case (Site ExperimentNumber) is
                    LogScaleClass_i) w_{i_1}
        (value of Velocity in case (Site ExperimentNumber)) is
                    VelocityClass_i) w_{i_2})
    ...etc. ...)) : ((1 1) (0 0))

where LogScaleClassi is a fuzzy set derived from the data, and w_{i_j} are importances. By examining the importances (as shown in Table 5.1), we see that LogScale is the most important feature in all cases. As a heuristic in determining the most important features, we also show in Table 5.1 the average importance of each feature, and calculate its average relative importance as follows.

Let the weight of the ith feature in the jth rule be w_{i_j}. The average importance of feature i is

$$\text{imp}_i = \frac{\sum_{j=1}^{m} W_{i_j}}{m}$$

where m is the number of rules, and the average relative importance is

$$\frac{\text{imp}_i \frac{1}{n}}{n},$$

where n is the number of features in the rule (*i.e.* the number of attributes considered).

Table 5.2: Importances of each attribute in predicting LogDisp.

	Log Disp Class1	Log Disp Class2	Log Disp Class3	Log Disp Class4	Log Disp Class5	Average importance	Average relative importance
LogScale	0.209	0.224	0.198	0.206	0.236	0.215	1.3607
Reliability	0.057	0.014	0.035	0.119	0.109	0.067	-0.266
Data Interpretation	0.119	0.111	0.128	0.126	0.108	0.118	0.3033
InputMethod	0.133	0.136	0.089	0.055	0.061	0.095	0.0396
Thickness	0.061	0.069	0.065	0.065	0.053	0.063	-0.312
Hydraulic Conductivity	0.049	0.057	0.056	0.060	0.047	0.054	-0.408
Hydraulic Transmissivity	0.051	0.059	0.059	0.060	0.047	0.055	-0.395
Effective Porosity	0.138	0.125	0.091	0.106	0.122	0.117	0.2829
Velocity	0.069	0.078	0.128	0.088	0.063	0.085	-0.061
Flow Configuration	0.112	0.060	0.053	0.092	0.119	0.087	-0.04
Monitoring	0.002	0.067	0.097	0.023	0.036	0.045	-0.505

If features were all equally important, the average relative importance of each feature would be zero. Using this formula we can see which features are more important, and which could perhaps be neglected. In this case, we obtain the results shown in Table 5.1. Clearly LogScale is the most important feature, followed by the data interpretation method and the effective porosity. In this case, it is not clear whether the high importance of the data interpretation method is valid, since it might be appropriate to define a similarity relation on the domain — for example, cluster together the two dimensional models, the three-dimensional models, etc.

Clearly, this approach may neglect a feature which is important in one rule, but not in any of the others, however it is adequate in this example. There are two indicators we use to assess the performance of rules. Consider a row in the database. We take the values for the attributes in the bodies of the rules, and find the support for each head, *i.e.* we obtain

LogDispClass1 : S1
LogDispClass2 : S2
LogDispClass3 : S3
LogDispClass4 : S4
LogDispClass5 : S5

Comparison of the support pairs $S_1 - S_5$ allows us to choose a class from this list; if the actual data value falls within this class, the example has been

Table 5.3: Importances of 3 selected attributes in predicting LogDisp.

	Log Disp Class1	Log Disp Class2	Log Disp Class3	Log Disp Class4	Log Disp Class5	Average importance	Average relative importance
LogScale	0.448	0.487	0.474	0.470	0.507	0.477	0.4313
Data Interpretation	0.256	0.241	0.307	0.287	0.231	0.265	-0.206
Effective Porosity	0.296	0.272	0.218	0.243	0.262	0.258	-0.225

correctly classified. Alternatively, we can calculate an *expected value* [22, 13] and compare it to the true value in the database. The average error in the expected value gives an indication of how well the rules perform. The evidential logic rules with all features classify 87% of cases correctly, and give an average error of 9.6% in predicting the point value.

Fuzzy logic rules were generated using the most important three features, LogScale, DataInterpretation and EffectivePorosity. These rules classify 88% of cases into the correct fuzzy category for LogDispersivity, and give an average error of only 7% when predicting a point value (see Figure 5.7, where results for LogScale with DataInterpretation and LogScale with Effective Porosity are also shown). We note that including extra features increases the accuracy of the rules in predicting values at the known points; however, it is not clear whether improved predictions will be achieved in between the known points. There is a danger that adding more features could actually *decrease* the generality of the rules, in much the same way as using a high order polynomial to fit data can give a less general model than using a straight line.

Creating evidential logic rules on the three most important attributes LogScale, EffectivePorosity, and DataInterpretation confirms the indication from Table 5.2, that LogScale is the single most important attribute (see Table 5.3).

Using just the LogScale attribute to build rules gives a prediction rate of 80% and an average error of 9% in the point value predicted by the rule (see Figure 5.7). This is in accordance with the expert view, that LogScale and LogDispersivity are roughly linearly related.

In Figure 5.7(a), all data points are included; in (b)–(d), the very uncertain point has been omitted from the plots for clarity, although it was used in the calculations.

(a) prediction on the basis of LogScale alone. Classification success 80%, average error in predicted point value 9%.

(b) prediction using LogScale and EffectivePorosity. Classification success 79%, average error in predicted point value 8%.

Table 5.4: Success of each attribute in predicting LogDisp.

Single attribute used in rule	%age of cases classified correctly	RMS error in predicted point value
LogScale	80	11
Reliability	75	14
DataInterpretation	85	14
InputMethod	74	18
Thickness	75	16
HydraulicConductivity	41	16
HydraulicTransmissivity	57	16
EffectivePorosity	39	17
Velocity	62	16
FlowConfiguration	81	15
Monitoring	68	16

(c) prediction using LogScale and DataInterpretation Classification success 85%, average error in predicted point value 8%.

(d) prediction using LogScale, EffectivePorosity and Data Interpretation. Classification success 88%, average error in predicted point value 7%.

It is noticeable that the curve predicted in (a) is considerably smoother than in the cases where two or more attributes are considered, since there is no third/fourth dimension to consider in (a). The extra degrees of freedom afforded by considering additional attributes may fit the existing data better, but it is arguable that the rule will be less general, *i.e.* that it will be less good at predicting unknown cases. This question is the subject of further investigation.

In the second approach to determining important features, we consider attributes one by one, and examine their efficiency in predicting the value of Log Dispersivity in each case. We can then look at combinations of features which make good predictions individually. This may introduce a decomposition error, as seen in the example of section 5.4.1, in that two features which make a good prediction in combination may perform badly individually. However we assume the individual performance is a good indicator, and we generate a set of rules of the form

((Predicted value for LogDisp in case (Site ExpNumber) is LogDispClass_i)
    (value of Attribute_j in case (Site ExpNumber) is
    Attribute_jClass_i)) : ((1 1) (0 0))

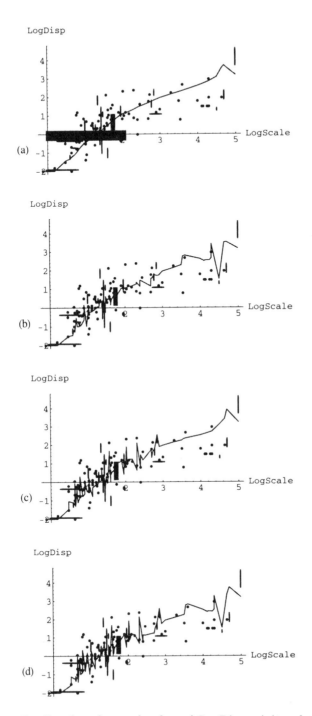

Figure 5.7: Predicted and actual value of LogDispersivity plotted against LogScale.

for each attribute, and consider the success of each rule set. This gives the results in Table 5.4.

Again, we see that LogScale is the single most important attribute, and that the method of data interpretation appears to have a significant influence on the value of the dispersivity. The flow configuration shows up as important in this method, and effective porosity is the worst predictor. Using this table as a guide, fuzzy logic rules were generated using the best three features, LogScale, DataInterpretation and FlowConfiguration. These rules classify 88% of cases into the correct fuzzy category for Log Dispersivity, and give an average error of only 7% when predicting a point value. This is the same level of performance as the rules created by considering feature importances.

Finally, we note that further tuning of the evidential rules may be possible by optimizing the form of the fuzzy set filter. This is the subject of ongoing research.

5.6   Summary

One of the longer term objectives of this work is to embed a subsystem into the Intelligent Manual [9] able to analyse and summarize mathematical or numerical models of engineering systems. Many databases and models may contain hidden redundancy, in that there are fuzzy underlying relations in the data. The Fuzzy Data Browser can assist a user in eliciting these relationships, and can provide summarized forms of complex models by means of a few rules. The fuzzy data browser has been used to extract fuzzy summarizing models in diverse areas, *e.g.*:

- monitoring data from an aircraft black-box flight recorder, where it is necessary to detect anomalous readings which can either indicate that a piece of equipment is malfunctioning, or that measurements are being reported incorrectly;

- classifying underwater sounds, where the browser must generate rules which can be understood by humans. This application has achieved very high success rates (up to 92%), outstripping the performance of a standard neural net package, yet retaining the transparency of the rule based approach. The neural net, in contrast, provides no explanation or insight into the classification process;

- generating rules for hand-written character recognition.

Further work is examining the possibility of using a toolkit where the user could design his own browser. Fuzzy object-oriented programming is a promising framework for this research.

References

[1] Baldwin J.F. (1981) Fuzzy Logic and Fuzzy Reasoning. In *Fuzzy Reasoning and its Applications*, Mamdani E.H. and Gaines B.R. (eds), Academic Press, 133–148.

[2] Baldwin J.F. (1991) Combining Evidences for Evidential Reasoning. *International Journal of Intelligent Systems* 6: 569–616.

[3] Baldwin J.F. (1991) A Calculus for Mass Assignments in Evidential Reasoning. In *Advances in the Dempster-Shafer Theory of Evidence* Fedrizzi M., Kacprzyk J. and Yager R.R. (eds).

[4] Baldwin J.F. (1992) The Management of Fuzzy and Probabilistic Uncertainties for Knowledge-Based Systems. In *Encyclopaedia of AI*, Shapiro (ed), Wiley (2nd ed) 528–37.

[5] Baldwin J.F. (1993) Evidential Support Logic, FRIL and Case Based Reasoning. *International Journal of Intelligent Systems* (8) 9: 939–961.

[6] Baldwin J.F. (1993) Fuzzy , Probabilistic and Evidential Reasoning in Fril. *Proc. 2nd IEEE International Conference on Fuzzy Systems*, San Francisco, CA, 459–464.

[7] Baldwin J.F. (1994) Soft Computing in Fril. *Proc. 4th Dortmund Fuzzy Days*, Springer Verlag.

[8] Baldwin J.F. and Martin T.P. (1991) An Abstract Mechanism for Handling Uncertainty. In *Uncertainty in Knowledge Bases*, Bouchon-Meunier B., Yager R.R. and Zadeh L.A. (eds), Springer Verlag, 126–135.

[9] Baldwin J.F. and Martin T.P. (1994) Fuzzifying a Target Motion Analysis Model Using Fril and Mathematica. *Proc. 3rd IEEE Int.Conf. on Fuzzy Systems*, 1171–1175.

[10] Baldwin J.F. and Martin T.P. (1995) Refining Knowledge from Uncertain Relations — a Fuzzy Data Browser based on Fuzzy Object-Oriented Programming in Fril. *Proc. 4th IEEE International Conference on Fuzzy Systems*, Yokohama, Japan, 27–34.

[11] Baldwin J.F., Martin T.P. and Pilsworth B.W. (1987) The Implementation of FPROLOG — a Fuzzy Prolog Interpreter. *Fuzzy Sets and Systems* 23: 119–129.

[12] Baldwin J.F., Martin T.P. and Pilsworth B.W. (1993) Fril: A Support Logic Programming System. In *AI and Computer Power: The Impact on Statistics*, Hand D. (ed), Chapman and Hall, 129–149.

[13] Baldwin J.F., Martin T.P. and Pilsworth B.W. (1995) *FRIL — Fuzzy and Evidential Reasoning in A.I.* Research Studies Press, John Wiley.

[14] Baldwin J.F. and Zhou S.Q. (1984) FRIL — A Fuzzy Relational Inference Language. *Fuzzy Sets & Syst.* 14: 155–74.

[15] Dubois D. and Prade H. (1991) Fuzzy sets in approximate reasoning 1 — inference with possibility distributions. *Fuzzy Sets and Systems* 40: 143–202.

[16] Gelhar L.W., Welty C. and Rehfeldt K.R. (1992) A Critical Review of Data in Field-Scale Dispersion in Aquifers. *Water Resources Research* (28) 7: 1955–1974.

[17] Warren D.H.D. (1983) *An Abstract Prolog Instruction Set.* SRI International, Menlo Park, CA.

[18] Wolfram S. (1991) *Mathematica: a system for doing mathematics by computer.* Addison Wesley.

[19] Sudkamp T. (1992) On Probability-Possibility Transformations. *Fuzzy Sets and Systems* 51: 73–81. North-Holland.

[20] Yager R.R. (1984) On Different Classes of Linguistic Variables defined via Fuzzy Subsets. *Kybernetes* 13: 103–110.

[21] Zadeh L.A. (1968) Probability Measures of Fuzzy Events. *Journal of Mathematical Analysis and Applications* 23: 421–427.

[22] Baldwin J.F. (1995) Knowledge from Data using Fril and Fuzzy Methods, *Fuzzy Logic*, John Wiley, 33–76.

6

Adaptive Fuzzy Systems for Modelling Static and Dynamic Processes

H. Bersini, V. Gorrini

6.1 Introduction

Since fuzzy systems and neural nets are easy to implement, robust, and can approximate to any degree any non-linear mapping, their use for adaptive control and system identification (mainly for non-linear processes) is expanding rapidly. Besides, it has been recently shown [16] that Sugeno's type of fuzzy systems provided certain restrictions on the architecture, and the operators used by the system were equivalent to radial-basis neural networks (RBF) (*i.e.* based on a local non-monotonous activation function like gaussian, rather than global monotonous ones like the sigmoid). In [4] we attempt a general characterization of the three systems MLP, RBF and FLC by emphasizing where they meet and where they part.

Many attempts have been proposed to realize classical controllers using fuzzy models. In this chapter, we first present a classical fuzzy controller and a fuzzy realization of a PID controller. These systems perform an highly non-linear control law and result to be quite robust, even in the case of noisy inputs. We have developed a gradient-based learning method (slightly improving on a method developed by Nomura *et al.* [13]) that performs optimization of given membership functions. This method, if the input space is uniformly covered, converges quickly to (local) minima and may be used in on-line optimization. Identification and control of non-linear processes still remains a difficult problem when the temporal structure is not exactly known. Dynamic processes of order superior to 1, *i.e.* mapping their output not only to the current input but also to previous inputs, are more difficult to approximate and to control than first order processes. We introduce a Recurrent Fuzzy System (RFS) that, in order to identify an unknown order model, uses recurrent rules (*i.e.* rules in which some output variables at time t become the input variables at time $t + 1$).

On the whole, fuzzy systems require two types of tuning which Lee [10] years ago already designated as structural and parametric tuning. The first one concerns the structure of the rules: the variables to account for, for each variable the partition of the universe of discourse, the number of rules and the conjunctions which constitute them. Once a satisfactory structure is available, a fine adjustment of the membership functions remains to be done. A lot of previous works aiming at the automatic discovery of a minimal adequate structure of fuzzy systems relied on a two steps strategy [2, 14, 20]: first the use of a clustering algorithm to situate a minimum number of membership functions, and subsequently an on-line optimization method for tuning the shape of these membership functions. Obviously this strategy requires possession from the very beginning of all the data upon which the global tuning of the fuzzy system is based. In some cases, like the on-line adaptation of fuzzy controllers, this requirement is beyond reach. Moreover, since we strongly believe that optimizing the structure alone does not make a lot of sense without accounting for the specific type of parametric learning responsible for the final performance, we favour a complete on-line and mixed strategy where the structure and parameters are simultaneously adjusted, then respecting the fact that the first adjustment is entirely dependent on the second. We know of few similar attempts [11], and a comparative analysis is under way. The last algorithm presented is developed in a biological spirit, and dedicated to the incremental building of fuzzy systems for function approximation. It is called EFUSS (Evolving Fuzzy Systems Structure), and aims at automatically and incrementally finding the minimal number of membership functions along with their appropriate shaping.

6.2   A Direct Adaptive Fuzzy Controller

6.2.1   Definition

In this section we concentrate on the simplest member of Sugeno's fuzzy systems given by the following type of mapping of $\mathbb{R}^n$ in $\mathbb{R}$, $Y = f_W(X)$: IF x_1 is A_1^k AND x_2 is $A_2^k K$ THEN y is γ^k where A_j^k represents the membership function coding for the linguistic term k associated to the variable j, and γ^k represents the crisp output of linguistic term k. Since in our gradient-based learning method we optimize the linguistic terms, and not blindly all the parameters, we assimilate the crisp output to a linguistic value (for instance, 1=*very small*, 3=*small*, 7=*big* and 10=*very big*). In all our approaches, we use isosceles triangles for membership functions (defined by the centre α_j^k and the base width β_j^k), algebraic product for intersection, and Sugeno's method for inference and defuzzification. Therefore, fuzzy output Y is derivable with respect to any controller parameter W (W = all $\alpha_j^k, \beta_j^k, \gamma^k, \forall j, k$) and $\partial Y / \partial W$ may be easily calculated. A gradient descent technique may be derived as follows. Let Z be the plant output and $E(Z)$ the error. We can write:

$$\Delta W = -\eta \frac{\partial E}{\partial W} = -\eta \frac{\partial E}{\partial Z} \frac{\partial Z}{\partial Y} \frac{\partial Y}{\partial W}.$$

What is unknown in the absence of an analytical model of the process is the Jacobian $\partial Z / \partial Y$. When using a gradient-based technique, a perfect knowledge of the Jacobian is not required [15]: since substituting $\partial Z / \partial Y$ by its sign ± 1 just alters the amplitude of the variation but not its direction, the gradient method can still give an adequate value of the control parameters.

If this fuzzy system is used as function approximator, the knowledge of the Jacobian is not required. In fact, as $E = E(Y)$ the adaptation rule will be:

$$\Delta W = -\eta \frac{\partial E}{\partial W} = -\eta \frac{\partial E}{\partial Y} \frac{\partial Y}{\partial W}.$$

6.2.2   A Classical Controller

In previous work involving one of the authors [15, 17], a simple Direct Adaptive Neural Controller (DANC) has been developed and tested on several toy problems on the cart-pole balancing problem. One and only one multilayer neural net was necessary, only the sign of the cart-pole Jacobian was known, and this was sufficent to achieve a very satisfactory control compared to numerous other neurocontrollers we were aware of, and to more conventional approaches like MRAC. In this section we describe in more detail the equivalent simple DAFC [3] similarly applied to the cart-pole problem (a problem frequently met in the NN and fuzzy control literature [1]). In our application, the objective of the control is to bring the cart-pole, whatever its initial conditions, to a stable equilibrium position with the four variables describing the system x, v, ϑ, ω, driven to 0.

The fuzzy controller is of Sugeno's basic type with four variables in the IF part (given by a 'fuzzy set' restricted to a single value), and the force to exert on the cart in the **THEN** part. An example of rule is: **IF** x is *large* **AND** v is *very small* **THEN** $F = 10N$ *(large)*.

The parameters to adjust are the membership functions coding the linguistic terms appearing in the premise of the rules, and the different possible values of the force appearing in the consequent part. With regard to the Jacobian, the signs of the four partial derivatives $\partial x / \partial F, \partial v / \partial F, \partial \vartheta / \partial F, \partial \omega / \partial F$ are all negative. The error is given by $E = 0.5(\alpha_1 x^2 + \alpha_2 v^2 + \alpha_3 \vartheta^2 + \alpha_4 \omega^2)$, where α_i weights the relative contribution of each variable. The global gradient method immediately follows:

$$\Delta W = -\eta \frac{\partial E}{\partial W} = -\eta \frac{\partial E}{\partial F} \frac{\partial F}{\partial W}$$

with $\partial E/\partial F$ easily obtained by substituting $\partial x/\partial F, \partial v/\partial F, \partial \vartheta/\partial F, \partial \omega/\partial F$ all by -1.

The results were extremely positive. For a large range of initial conditions, including extreme conditions, the equilibrium position was reached after an average of 5 falls and of 2500 learning steps. For these experiments, and to simplify the comparison with the results of the equivalent neural controller for the same problem, the rules were generated at random, like a synaptic matrix. More precisely, any initial set of rules contains 30 rules. Each variable is associated to a linguistic term taken randomly among seven possibilities. On the whole the learning was faster than for the neural equivalent, and the final controller tolerated more variation on the initial conditions and was more robust than the neural controller, confirming the results exposed by Kosko [9] for another application. Furthermore, the equilibrium position could even be reached without a fall, provided the initial set of rules was no longer taken randomly but on the basis on some simple common sense. Normally, an equivalent *a priori* would be difficult to install in the synaptic matrix of a neural net. This last point illustrates a very crucial issue and a definite advantage of fuzzy controllers with respect to neural controllers.

6.2.3   A PID Controller

Fuzzy PID is one of the most popular fields of investigation in the fuzzy control community, and researchers are trying to understand better the kind of nonlinear extrapolation that the fuzzification of a classical PID can provide [6]. Hence we decided to extend our adaptive mechanism to a fuzzy PID [3]. By defining:

$$
\begin{aligned}
e(t) &= y(t) - y^\circ(t) \quad \text{(with } y^\circ(t) \text{ the desired output for the process)} \\
\delta e(t) &= e(t) - e(t-1) = y(t) - y(t-1) - y^\circ(t) + y^\circ(t-1) \\
\delta\delta e(t) &= \delta e(t) - \delta e(t-1) = y(t) - 2y(t-1) + y(t-2) - y^\circ(t) \\
&\quad + 2y^\circ(t-1) - y^\circ(t-2) \\
u(t) &= u(t-1) + \delta u(t)
\end{aligned}
$$

a fuzzy PID is a particular member of Sugeno's systems composed of a bunch of fuzzy rules similar to IF e is NL AND δe is PM AND $\delta\delta e$ is ZE THEN δu =NL.

Although this controller is different from the previous one due to the nature of its inputs (error and error derivatives) and outputs (increment of control), the adaptive mechanism is basically the same. We tested this adaptive method on the process $y(t) = ay(t-1)\sin(y(t-2)) + \cos(ku(t-1))$. For such a process, the fuzzy PID is not easy to tune manually, and an adaptive method is preferred for its automatic discovery capabilities. Our fuzzy controller contains 27 rules, generated in a way that tries to keep the proportionality inherent to the PID structure [6]. The final fuzzy PID found by the optimization algorithm leads to an efficient and quite robust controller.

6.2.4   A Recurrent Fuzzy System

Identification and control of non-linear processes still remains a difficult problem when the temporal structure is not known exactly. Dynamic processes of order superior to 1, *i.e.* mapping their output not only to the current input but also to previous inputs, are more difficult to approximate and to control than first order processes.

A recurrent fuzzy systems (RFS) [7] is a fuzzy system with inferential chaining in which some output variables at time t become the input variables at time $t+1$; an approach very typical of fuzzy reasoning or fuzzy expert system, but completely infrequent in fuzzy control where one shot input/output structures are commonplace. This connection between fuzzy reasoning and fuzzy control might allow a better overlapping than the one existing today between developments in fuzzy logic and fuzzy controllers. This specific approach has become very popular in the connectionist community, and today the use of recurrent neural networks both for process identification and control is rapidly expanding [12, 21].

A first order RFS can only be used for approximating a first order process *i.e.* $x(t) = f(x(t-1), u(t-1))$. However a richer structure, including internal variables, can still allow a first-order structure to be used for approximating a process of order superior to 1, *i.e.* $x(t) = f(x(t-1), x(t-2), \ldots, x(t-n), u(t-1), u(t-2), \ldots, u(t-m))$. We restrict to RFS with a single-input-two-internal-variables-single-output structure. The whole system can be split in two subsystems, each one containing a certain number of fuzzy rules: a first one (f/g) which captures the temporal dynamics of the two internal variables s and p, and a second one (h) which gives the output x. The resulting complete system $(f/g+h)$ can be represented in this economical: way: a first subsystem:

$$h: \quad x(t) = h\left(x(t-1), s(t), p(t), u(t-1); \vec{\alpha}, \vec{\beta}, \vec{\gamma}\right)$$

containing rules of the form:

IF $x(t-1)$ is $A(\alpha_j^k, \beta_j^k)$ **AND** $s(t)$ is $A(\alpha_j^k, \beta_j^k)$ **AND** $p(t)$ is $A(\alpha_j^k, \beta_j^k)$

AND $u(t-1)$ is $A(\alpha_j^k, \beta_j^k)$ **THEN** $x(t)$ is γ^k

and a second one:

$$f/g: \begin{cases} s(t) & = & f(x(t-1), s(t-1), p(t-1), u(t-1); \vec{\vartheta}, \vec{\varphi}, \vec{\chi}) \\ p(t) & = & g(x(t-1), s(t-1), p(t-1), u(t-1); \vec{\vartheta}, \vec{\varphi}, \vec{\xi}) \end{cases}$$

containing rules of the form:

IF $x(t-1)$ is $A(\vartheta_j^k, \varphi_j^k)$ **AND** $s(t-1)$ is $A(\vartheta_j^k, \varphi_j^k)$ **AND** $p(t-1)$ is $A(\vartheta_j^k, \varphi_j^k)$

AND $u(t-1)$ is $A(\vartheta_j^k, \varphi_j^k)$ **THEN** $s(t)$ is χ^k **AND** $p(t)$ is ξ^k

where α, β, γ are vectors of parameters for the positions, the bases and the outputs of the rules constituting the subsystem h and $\vartheta, \varphi, \chi, \xi$ are vectors of parameters for the subsystem f/g. Since only the output of f and g contain different variables, only the two vectors of parameters shaping the output of the rules that constitute the subsystems f and g, $i.e.$ χ and ξ are different.

Once we have defined the square error E at time t we try to adjust all parameters using a gradient descent method (this approach is very similar to William and Zipser's algorithm for recurrent neural networks [21]). Derivatives of E with respect to different parameters $(\alpha \dots \xi)$ are calculated in recurrent way [7].

We used the RFS shown in section 6.2 to approximate a third order non-linear system:

$$y(t) = \frac{y(t-1)y(t-2)y(t-3)[y(t-3)-1]u(t-2) + u(t-1)}{1 + y^2(t-2) + y^2(t-3)}.$$

Each fuzzy sub-system is defined using 30 fuzzy rules randomly generated and nine linguistic terms for each variable. The initial membership functions are settled so as to uniformly cover their space state. Our results appear very promising compared to those given by a recurrent neural network used for the identification of the same process [12].

6.3   EFUSS — A Self-Structuring Fuzzy Systems for Function Approximation

In this section we will concentrate on the simple member of TSF given by the following type of mapping of $\mathbb{R}$ in $\mathbb{R}$, $y = f(u)$: IF u is A^k THEN y is B^k where the output B^k represents the linear response $\gamma_k + \delta_k u$ (and not a singleton as in section 6.2). In this section, a brief description of the algorithm together with the application and experimental results presented will be restricted to the approximation of monovariable functions. We are currently working on a extension of the same basic principles underlying this algorithm to the approximation of multivariable functions.

The parametric learning algorithm described above allows the system to tune gaussian centres and width (α, β) but also the output parameters (γ, δ). Now clearly some zones need a lot more membership functions than others due to the stronger nonlinearity of the function profile over these specific zones. For instance in Figure 6.1, the right part should certainly be given more membership functions than the left part.

The main ideas underlying EFUSS [8] are: first to observe the oscillatory tendency of the parameters defining the output part of the TSF fuzzy rules; then detecting the most oscillatory one; and finally supplying the zone covered by the input of this strongly oscillating rule with a complementary fuzzy rule.

In a preliminary phase, the gradient method applied on fuzzy approximators with too many or too few membership functions was tested to understand

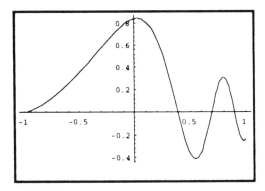

Figure 6.1:

better the dynamics of the parameters for those cases. We plotted the fuzzy rules' output parameters as a function of time. The curves turned out to be quite complex showing initialization periods, a long period of high amplitude oscillations and other complex features. Eventually we came out with a relevant criterion to detect the successful and failing rules: the output parameters' high frequencies magnitude *HF*. We saw a high correlation between gaussian location difficulty, the strong oscillatory tendency of their associated output and their respective *HF*. It seems natural then to focus on the rule showing the highest *HF* in order to duplicate it. When we tried to approximate functions with enough well positioned gaussians, their respective *HF*s were very low, which can be taken as a validation of our whole approach. Moreover, the use of *HF* leads to an obvious stopping criterion for our algorithm: if the *HF* remains below the *duplication threshold* the incremental building can be stopped. By now we have defined a criterion to locate which membership function ought to be duplicated and when to stop adding new gaussians, but what the algorithm still doesn't know is when to perform a duplication. The best way is to wait for the fuzzy system to stabilize with its current parameters, namely to do the best it can in its current form and to stop changing the current approximation error. If the error derivative is under the *stagnancy threshold* the system is considered to be stagnant and the algorithm starts, from this moment, to compute *HF* for all rules. After the *averaging time*, it determines which rule has the highest *HF*. When a new gaussian is added, the centres of the new and old ones are given an equal value: the old centre, *i.e.* the two gaussians, are superimposed. The gaussian base width is given a fixed value: the *initial base width*. Regarding the output parameters, during the *averaging time* their average values and standard deviations are computed. The old and new rules received as value for their output parameters the average value plus the standard deviation for the first one and less the standard deviation for the second one.

To test EFUSS we tried to approximate a monovariable function $x = f(u)$. No attempt has been made yet on multivariable functions, but nothing in our methodology should prevent its extension for multivariable functions. This extension is in progress. The function (see Figure 6.1) is:

$$x = \frac{\sin(u+1)^{3.5}}{(u+1)^2} \qquad u \in (-1, +1).$$

As you can see in Figure 6.1, the right side is more complex and, in principle, should demand more gaussians to perform a satisfactory approximation. The full process was initiated with three gaussians homogeneously distributed on the input space, and it stopped by itself following the addition of five gaussians (so eight gaussians as a whole). Regarding the approximations results, EFUSS obtained better results (*i.e.* smaller final error and faster convergence) than fuzzy systems with homogeneously distributed gaussians

6.4   Conclusions

The main point developed in this chapter is the ease of importing methods and ideas that have appeared in the connectionist community for control problems, as soon as the fuzzy controller is supplied with gradient-based automatic tuning of its parameters. We have presented two different DAFCs, the second one coupling a fuzzy PID with an automatic adjustment mechanism, and have shown the good performance obtained when applying these two DAFCs to simple control problems. The signs of the Jacobian represents the minimal knowledge required about the plant. This idea fits in perfectly with the philosophy of fuzzy logic, where the world is represented in a qualitative, linguistic way, avoiding very precise numerical expression.

The use of recurrent fuzzy systems for the identification and control of an n-th order process creates a connection between fuzzy reasoning and the type of systems used for fuzzy control: both systems run by means of inferential chaining, although in the case of RFS, the chaining is of length 2. It is well known that when chaining fuzzy rules, new problems appear (according to researchers in fuzzy logic, the only interesting ones) if the output of the first rules is not defuzzified first to become the input of the second chained rule: this problem has been called fuzzy modus ponens, and has been the subject of intensive work in the fuzzy logic community [19]. In the work presented in this chapter, this problem was eliminated by using Sugeno's inferential techniques, in which no real defuzzification is necessary since nothing is kept fuzzy at the output level.

Although we still haven't developed the complete version of EFUSS for multivariables functions, we are confident that this could be most appropriate for the complete and autonomous discovery of performant fuzzy systems. First of all, this algorithm does not separate the structural adaptation from the parametric one, like a large number of two-step adaptation methods. Then the structural changes really aim at providing compensatory help for parametric learning. Secondly, and indrawing inspiration from biological systems, the structural changes are entirely guided by the endogenous dynamics of the system, *i.e.* it is the system which introspectively calls for internal changes, independent of exogenous factors associated with the specific problem. Thirdly, a very similar type of strategy called EMANN [18] has been applied with success for the automatic discovery of neural net classifiers, and we are encouraged to think that respecting such simple principles like slower structural changes, guided by the endogenous dynamics and compensating for the weakest parts, might be generalizable to all engineering systems characterized by double level plasticity: structural and parametric.

Acknowledgement

This work has been parially supported by a fellowship granted to Vittorio Gorrini for the years 1994–95 by the Université Libre de Bruxelles.

References

[1] Barto A., Sutton R. and Anderson C. (1983) Neuronlike adaptive elements that can solve difficult learning control problems. *IEEE Transactions on Systems, Man and Cybernetics* (13) 5:.

[2] Berenji H.R. and Khedkar P.S. (1993) Clustering in Product Space for Fuzzy Inference. In *Proceedings of the Second IEEE International Conference on Fuzzy Systems* 1402–1407.

[3] Bersini H. and Gorrini V. (1993) FUNNY (FUzzY or Neural Net) Methods for Adaptive Process Control. In *Proceedings of the first European Congress on Fuzzy and Intelligent Technologies* 55–61.

[4] Bersini H. and Gorrini V. (1994) MLP, RBF, FLC: What's the Difference? In *Proceedings of the second European Congress on Fuzzy and Intelligent Technologies.*

[5] Bersini H., Nordvik J-P. and Bonarini A. (1993) A Simple Direct Adaptive Fuzzy Controller Derived from its Neural Equivalent. In *Proceedings of the Second IEEE International Conference on Fuzzy Systems* 345–350.

[6] Foulloy L. and Galichet S. (1992) Controlleurs Flous: représentation, équivalences et études comparatives. Rapport LAMII 92-4.

[7] Gorrini V. and Bersini H. (1994) Recurrent Fuzzy System. In *Proceedings of the Third IEEE International Conference on Fuzzy Systems* 193–198.

[8] Gorrini V., Bersini H. and Salomé T. (1995) Self-Structuring Fuzzy Systems for Function Approximation. *Proceedings of the FUZZ-IEEE '95 Conference.*

[9] Kong S.G. and Kosko B. (1992) Adaptive Fuzzy Systems for Backing up a Truck-and-Trailer. *IEEE Transactions on Neural Networks* (3) 2: March, 211–223.

[10] Lee (1990) Fuzzy logic in control systems: Fuzzy logic controller — Parts I, II. *IEEE Transactions on. Systems, Man and Cybernetics* (20) 2: 7–31.

[11] Lin C.T. and Lee C.S.G. (1992) Real-Time Supervised Structure/Parameter Learning for Fuzzy Neural Network. In *Proceedings of the First IEEE International Conference on Fuzzy Systems* 1283–1291.

[12] Narendra K.S. and Parthasarathy K. (1992) Neural Networks and Dynamical Systems. *International Journal of Approximate Reasoning* 6: 109–131.

[13] Nomura H., Hayashi I. and Wakami N. (1992) A learning method of fuzzy inference rules by descent method. In *Proceedings of the First IEEE International Conference on Fuzzy Systems* 203–210.

[14] Pacini J.P. and Kosko B. (1993) Adaptive Fuzzy Frequency Hopping System. In *Proceedings of the Second IEEE International Conference on Fuzzy Systems* 1113–1118.

[15] Renders J.M., Bersini H. Saerens M. (1993) Adaptive NeuroControl: How Simple and BlackBox can it be? In *Proceedings of the 10th International Conference on Machine Learning.*

[16] Roger Jang J.S. and Sun C.-T. (1993) Functional equivalence between radial basis function networks and fuzzy inference systems. *IEEE Transactions on Neural Networks* (4) 1: 156–159.

[17] Saerens M., Soquet A., Renders J.M. and Bersini H. (1992) Some Preliminary Comparisons between a Neural Adaptive Controller and a Model Reference Adaptive Controller. Bekey G. and Goldberg K. (eds) *Neural Networks in Robotics.* Kluwer Academic, 131–146.

[18] Salomé T. and Bersini H. (1994) An Algorithm for Self-Structuring Neural Net Classifiers. In *Proceedings of the 1994 IEEE Conference on Neural Networks* 1307–1312.

[19] Ph. Smets (1991) Implication and Modus Ponens in Fuzzy Logic. Goodman I.R., Gupta M.M., Nguyen H.T. and Rogers G.S. (eds) *Conditional Logic in Expert Systems*, Elsevier Science, 235–268.

[20] Sugeno (1993) A Fuzzy-Logic-Based Approach to Qualitative Modeling. In *IEEE Transactions on Fuzzy Systems* (1) 1: 404–435.

[21] Williams R. and Zipser D. (1989) A learning algorithm for continually running fully recurrent neural networks. *Neural Computation* 1: 270–280.

7

Fuzzy Logic in Diagnosis: Possibilistic Networks

M. Ulieru

7.1   Introduction

A *diagnostic problem* is understood as being the task which aims to *explain a system's misbehaviour* by analysing the relevant features (symptoms) which characterize it. As a result of this explanation, the misbehaviour is usually attributed to (labelled as) a *fault*, if technical systems are involved, or *a disease* in the case of biological ones. Both in the case of medical as well as technical diagnosis, the misbehaviour (illness/malfunction) usually doesn't show up obviously, but its *characteristic features* must somehow be *extracted* from the diagnosed system. In general, the feature extraction is made by direct observations, measurements or special tests applied to the investigated system. The results of these investigations have either a numerical form or they are expressed linguistically. This brings the necessity to *integrate heterogenous linguistic and numerical information*.

Once the relevant symptoms have been extracted, the most probable explanation is searched. The human diagnostician tries to find this explanation by using his previous experience and knowledge (involving a very good understanding of the laws governing the system's behaviour.) This diagnostic information, heterogenous in its nature (analytical, heuristic, statistical), is *structured* in a manner which underlines the *interdependencies* between *symptoms* (as manifestations of a system's misbehaviour) and their *explanations* (formulated as diagnoses). The structured information is called *a diagnostic model*.

Solving the diagnostic problem is one of the tasks which mostly requires human judgement. Therefore, it is extremely difficult to embed it into an

FUZZY LOGIC
Editor J. F. Baldwin   ©1996 John Wiley & Sons Ltd.

adequate computational frame. In our opinion this can be done adequately only by emulating the diagnostician's way of thinking and understanding of the diagnostic phenomenology.

The behaviour of the diagnosed system is a continuous dynamic flow which can only be explained in terms of the *propagation of events* in the environment (*i.e.* changes of the states or configurations of objects) after decomposition into discrete events. The explanation depends on this decomposition and search for unusual conditions and events. The propagation of events from the fault/disease to its manifestations (symptoms) is a very *complex phenomenology embedded in an appropriate way* in the diagnostician's mind, in the form of *intercorrelated associations* by which the correlated events are approximately connected, and/or logical connectives which in turn themselves have an approximate rather than precise meaning (something like an ANDOR which can be adequately modelled via a fuzzy connective, *e.g.* a parametrized operator [33]). Concluding, to cope with the complexity of the diagnostic phenomenology, the diagnostician builds (based on his previous experience) a mental model which approximately structures the correlations of events as a soft network consisting both of approximate links and approximate nodes.

To find out the explanation, some kind of *judgement* or *reasoning* about the diagnosed system is made through which the detected symptoms are analyzed by *analogy* with previous knowledge (encoded in the mental diagnostic model). On a global scale, the diagnostician evaluates the extent to which his mental model encoding the diagnostic knowledge is *matched* by the new information brought via the symptoms detected. The explanation is searched by attempting to *fit* the available data into the existing mental model [1]. This sort of *analogical reasoning* based on *approximate matching* of the fuzzy models with the current information is fairly emulated by *possibilistic reasoning* [31, 3] — the reasoning system based on fuzzy logic. Therefore, the involvement of fuzzy set theory when attempting to embed diagnostic reasoning within computers appears natural [29] (Figure 7.1).

The performance of continuous dynamic systems is in many cases characterized by a gradual transition from the range of acceptable behaviours to that of unacceptable ones [12]. This defines an *unsharp boundary* between *faulty/non-faulty* behaviour which has to be adaquately managed, especially when the fault develops *incipiently*. In this case the fault shows up *gradually*, via *dynamic symptoms* developing continuously in time as increasing or decreasing *process variables* (measurable signals, parameters, states). Sometimes these symptoms occur intermittently. An important role in incipient fault detection is played by those symptoms detected by *subjective human perceptions* (*e.g.* a *strange sound* or the *smell of burning*) which cannot be monitored but can be integrated with the process variables as fuzzy subsets [29].

In the following, *approximate reasoning on possibilistic networks* is presented as a tool for *diagnosing complex faults in continuous dynamic systems*.

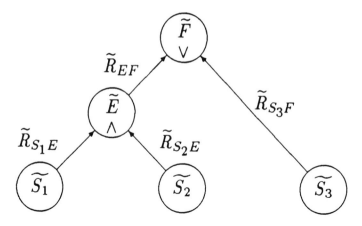

Figure 7.1: Strategy of diagnosis with fuzzy logic.

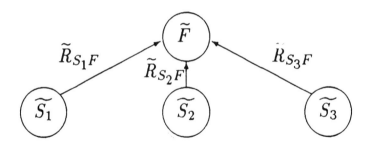

Figure 7.2: Soft diagnostic network.

7.2 The Possibilistic Network as a Diagnostic Model

From the above considerations it results that both the diagnostic information itself as well as its structure as diagnostic knowledge in the diagnostician's mind are approximate, rather than exact, this approximation being triggered by the complexity of the diagnostic phenomenology. This approximation can be fairly modelled as a *soft network* [2], linking via *approximate connections* events *approximately described* in a linguistic way. This brings us straight-forwardly to a network-like representation where the nodes-variables (representing the events) are linguistically described as fuzzy subsets and the links as approximate relations between these approximate nodes, defined as fuzzy relations (Figure 7.2) [25].

By enlarging the space of the variables attached to the nodes so that they

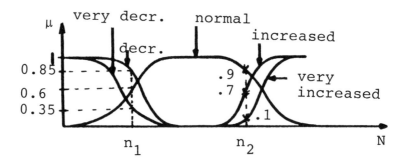

Figure 7.3: Possibilistic dimensions of N.

can take a range of fuzzy-set values in the universe represented by each node, say

$$\tilde{N}^k : N \to [0, 1], \ k = 1 \div r \tag{7.1}$$

the knowledge borne by the network is enriched and refined, covering several possible configurations of the domain modelled. The basic structure of the rules modelled by the links is kept, but a whole spectra of k-possible causal configurations, $k = 1 \div r$, are defined — one for each causal combination of distinct fuzzy-set values assigned to the network's nodes as *possibility distributions*

$$\Pi_N^k \equiv \tilde{N}^k. \tag{7.2}$$

The *states of the fuzzy variables* are identified with *membership values* $\mu_N^K(n)$, $n \in N$. A variable can simultaneously be in *several states*, with different *degrees*, implicitly determined by (and identified with) the degrees of belonging to the fuzzy subsets $\tilde{N}^k$. By this, several configurations of the fuzzy variable are accomplished with different degrees. The *number of states with nonzero degrees* depends on how many fuzzy subsets overlap in the respective states. For example, in Figure 7.3 value n_1 of the fuzzy variable N is simultaneously in states *vary decreased* with 0.35 degree, *decreased* with 0.85 degree and normal with 0.6 degree. Analogously, state n_2 is *very increased* with 0.1, *increased* with 0.7 and normal with 0.9. Incomplete knowledge about the state of the object is implicitly encoded within the definition of the fuzzy subsets regarded as possibility distributions, with the possibility distribution functions identified with the respective membership functions (7.2):

$$\pi^k(n) \equiv \mu_N^k(n). \tag{7.3}$$

The links associate the fuzzy variables (*e.g.* $F - E$ in Figure 7.2) as *joint possibility distributions*, say

$$\prod_{(C \times E)}^{k} = \tilde{F}^k \times \tilde{E}^k, \ k = 1 \div r. \tag{7.4}$$

The way in which the values of the fuzzy variables are associated is *contextually dependent*. However, for clarity of presentation, in the following it is assumed that all the fuzzy variables defining nodes that communicate through a path are associated through their k-value fuzzy set, as above.

Each parent node associated to its direct successors E_j, $j = 1 \div m$, is globally modelled as a *multi-antecedent fuzzy IF-THEN rule*, with a spectra of r configurations:

$$\mathtt{IF}\{\mathtt{OR}[\underset{j=1}{\overset{p}{\mathtt{AND}}} < E_j^k \text{ is } \tilde{E}_j^k >, \ldots, \underset{j=q}{\overset{m}{\mathtt{AND}}} < E_j^k \text{ is } \tilde{E}_j^k >]\} \tag{7.5}$$

$$\mathtt{THEN} \ < F^k \text{ is } \tilde{F}^k >, \ k = 1 \div r$$

represented as a joint possibility distribution:

$$\prod_{(C \times E_1 \times \cdots \times E_m)}^{k} \equiv U[\bigcap_{j=1}^{p} \tilde{E}_j^k, \ldots, \bigcap_{j=q}^{m} \tilde{E}_j^k] \times \tilde{F}^k \tag{7.6}$$

with the possibility distribution function implemented by the *intersection* of the effects combination with their parent node:

$$\pi^k(f, e_1, \ldots, e_m) = T\{S''[T''_{j=1}^{p} \mu^k(e_j), \ldots, T''_{j=q}^{m} \mu^k(e_j)], \mu^k(f)\}, \tag{7.7}$$

where T is a fuzzy intersection operator and S''/T'' a pair of fuzzy **AND/OR** connectives mathematically modelled by triangular norms [5]. Broadly, possibilistic networks are *multivariable fuzzy systems* characterized by the following:

- The relaxed associative frame allows the reasoning to be directed towards the explanation. This ensures a goal-oriented approach, which emulates the *human problem-solving* approach. Indeed, a human problem-solver unconsciously begins the problem-solving process with the consideration of the desired output of his actions. This *pursuit* of a goal guides both *information collection* and *hypothesis formation* [1].

- Possibilistic networks are *soft knowledge networks*. Their structure is covered with nuanced spectra of soft information represented in a (soft) logical manner (the connectives are fuzzy, with the meaning placed somewhere between the crisp **AND** and **OR**) either as uni- or as multi-antecedent **IF-THEN** rule. Therefore, they are quite fair models for human mental representation of the diagnostic information. Indeed, mental models are mostly inaccurate, based of soft associations of concepts [1] (recall as well the remarks made in the introduction).

In the following a reasoning frame for diagnosis by possibilistic networks is proposed [27] which emulates a human diagnostician's reasoning.

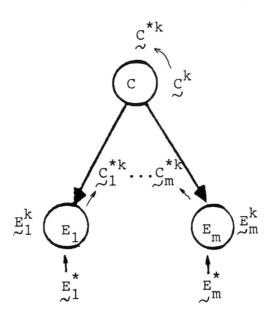

Figure 7.4: Possibilistic reasoning for configuration k.

7.3 The Particularities of Possibilistic Inference in Diagnosis

7.3.1 General Case

For the simple example in Figure 7.4 the *possibilistic inference* works in the following steps:

7.3.1.1

The *observed* events (symptoms) bring *new evidence* in the form of the *possibility distributions* $\tilde{E}_j^*$, $j = 1 \div m$ (inputs).

7.3.1.2

Global approximate reasoning strategy for configuration k:

$$
\begin{array}{ll}
\text{FACTS} & \tilde{E}_1^*, \ldots, \tilde{E}_m^* \hspace{2cm} (7.8) \\
\text{RULE} & \prod_{(F \times E_1 \times \cdots \times E_m)}^k \equiv U[\bigcap_{j=1}^p \tilde{E}_j^k, \ldots, \bigcap_{j=q}^m \tilde{E}_j^k] \times \tilde{F}^k \\
\hline
\text{CONCEQUENCE} & \tilde{F}^{*k} = (\tilde{E}_1^*, \ldots, \tilde{E}_m^*) \circ \prod_{F \times E_1 \times \cdots \times E_m}^k
\end{array}
$$

7.3.1.3

Decompositional inference. According to Kosko's *Decomposition Theorem* [10] (see also [29]), each *k-global inference* (7.9) for a *rule configuration at explanatory node F* can be split into several *simple k-local inferences on each link* as follows:

$$\tilde{F}^{*k} = (\tilde{E}_1^*, \ldots, \tilde{E}_m^*) \circ U[\bigcap_{j=1}^{p} \tilde{E}_j^k, \ldots, \bigcap_{j=q}^{m} \tilde{E}_j^k] \times \tilde{F}^k \tag{7.9}$$

$$= U[\bigcap_{j=1}^{p} \tilde{E}_j^* \circ (\tilde{E}_j^k \times \tilde{F}^k), \ldots, \bigcap_{j=q}^{m} \tilde{E}_j^* \circ (\tilde{E}_j^k \times \tilde{F}^k)].$$

7.3.1.4

The *k-explanatory evidence* at node F is:

$$\mu^{*k}(f)$$
$$= S''\{T''_{j=1}^{p} \sup_{e} T'[\mu_j^*(e), \mu_{R_j}^k(e, f)], \ldots, T''_{j=q}^{m} \sup_{e} T'[\mu_j^*(e), \mu_{R_j}^k(e, f)]\}$$
$$\hat{=} S''\{T''_{j=1}^{p} \mu_j^{*k}(f), \ldots, T''_{j=q}^{m} \mu_j^{*k}(f)\}, \tag{7.10}$$

where $T, T''/S''$ is as in (7.7) and T' is a triangular norm involved in the compositional rule of inference.

7.3.1.5

Inference steps.

a) Consider the terms $\mu_j^{*k}(f)$ in (7.10). If the norms $T \equiv T'$ and *lower semicontinuous* [5], *i.e.* satisfying:

$$\sup_{s} T[u, \mu(s)] = T[u, \sup_{s} \mu(s)], \quad \text{any} \quad u \in [0, 1] \tag{7.11}$$

(*e.g.* MIN or product), then:

$$\mu_j^{*k}(f) = \sup_{e} T[\mu_j^*(e), T\{\mu_j^k(e), \mu^k(f)\}] = T[\sup_{e} T\{\mu_j^*(e), \mu_j^k(e)\}, \mu^k(f)] \tag{7.12}$$

and each simple/local *k*-inference on a link is done in two steps:

- Compute the *degrees of similarity* α_j^k of the possibility distribution of each observed effect $\tilde{E}_j^*$ to the *prior possibility distributions* $\tilde{E}_j^k$ assigned to the effect nodes within the *k*-possible configuration:

$$\alpha_j^k = \sup_{e} T\{\mu_j^*(e), \mu_j^k(e)\}. \tag{7.13}$$

Note that α_j^k, $j = 1 \div m$, are numbers in the unit interval, representing the *k-possibility measures* [31] of the observed effects $\tilde{E}_j^*$.

- Update the k-prior possibility distribution $\tilde{F}^k$ of the explanatory fuzzy variable to the possibility measure α_j^k of the observed effect $\tilde{E}_j^*$:

$$\mu_j^{*k}(f) = T[\alpha_j^k, \mu^k(f)] \tag{7.14}$$

b) Introducing (7.14) in (7.10):

$$\mu^{*k}(f) = S''\{\underset{j=1}{\overset{p}{T''}} T[\mu_j^k, \mu^k(f)], \ldots, \underset{j=q}{\overset{m}{T''}} T[\alpha_j^k, \mu^k(f)]\} \tag{7.15}$$

$$= T\{S''[\underset{j=1}{\overset{p}{T''}} \alpha_j^k, \ldots, \underset{j=q}{\overset{m}{T''}} \alpha_j^k]\mu^k(f)\} \hat{=} T\{\alpha^k, \mu^k(f)\}$$

it comes out that, under this implementation, the *k-explanatory possibility* $\mu^{*k}(c)$ is obtained by weighting (via rule's norm T) the *k-prior possibility* describing the explanatory variable F with the aggregated value α^k of the possibility measures α_j^k of the observed effects $\tilde{E}_j^*$.

The same result is obtained if, for example, in (7.10) $T \neq T'$ (and both lower semicontinuous!), but the cause variable is implemented by *singletons*.

c) The global contribution of each rule-configuration is gathered into a *general posibilistic explanatory information*:

$$\tilde{F}^* = \bigcup_{k=1}^{r} \tilde{F}^{*k}. \tag{7.16}$$

d) Extract from the global $\tilde{F}^*$ the *relevant information* as the *degree of evidence of the explanation*, given by its *possibility measure*:

$$PM_C = \sup_c \mu_C^*(c). \tag{7.17}$$

The *size* of F, in case that it is quantifiable (*e.g.* a physical quantity in case of technical diagnosis), can be estimated as

$$f^* = \mathrm{DEF}(\tilde{F}*), \tag{7.18}$$

where DEF symbolizes a defuzzification formula and f^* is regarded as *the most possible state c^** explaining the currently observed effects.

7.3.2 Significant Particularisations

a) If in (7.16) $T = MIN$/product, the k-explanation is an α-cut/product cutting/weighting the *a priori* distribution assigned to the explanation $\tilde{F}$ at/by the aggregated possibiiities α^k.

b) If $\tilde{F}^k$ are *singletons* (*i.e.* if $\mu_F(f) = 1$ and $\mu_F^k(f') = 0$, any $f' \neq f$) then the *k-explanatory evidence* equals α^k:

$$\mu_j^{*k}(f) = \sup_e T\{\mu_j^*(e), T[\mu_j^k(e), 1]\} = \sup_e T\{\mu_j^*(e), \mu_j^k(e)\} \hat{=} \alpha$$

$$= T\{\alpha, 1\} = T\{\alpha, \mu_F^k(f)\} \qquad \text{q.e.d.} \qquad (7.19)$$

c) If, to conditions a) and b), we add:

- $r = 1$ (*i.e. one* configuration)
- $T''/S'' \equiv$ product/bounded sum
- E_j are *singletons*
- $\alpha_j = P^*(E_j)$,

then the inference on possibilistic networks gives the same result as the Bayesian one under the statistical independence of variables [24].

7.4   Illustration on a Simple Example

7.4.1   General Case

Consider the example of tree-like network in Figure 7.2. As *inputs* we have the detected symptoms s_1, s_2, s_3 either as crisp values (singletons) or represented as fuzzy numbers.

A. Inference at node E (**AND** node)

$$\tilde{E}_k^* = [\tilde{S}_1^* \circ \tilde{R}_{S_1 \times E}^k] \bigcap [\tilde{S}_2^* \circ \tilde{R}_{S_2 \times E}^k], \quad k = 1 \div r \qquad (7.20)$$

$$\mu_{E_k}^*(e) = \sup_{s \in S_1} T'\{\mu_{S_1}^*(s), \mu_{R_{1E}} k(s, e)\} \wedge \sup_{s \in S_2} T'\{\mu_{S_2}^*(s), \mu_{R_{2E}} k(s, e)\}. \qquad (7.21)$$

Each inference at node E (**AND** node) turns into:

$$\mu_E^*(e)$$
$$= \sup_{s \in S_1} T'\{\mu_{s_1}^*(s), T[\mu_{S_1}(s), \mu_E(e)]\} \wedge \sup_{s \in S_2} T'\{\mu_{S_2}^*(s), \mu_E(e)\}$$
$$= T\{\sup_{s \in S_1} T'[\mu_{S_1}^*(s), \mu_S(s)], \mu_E(e)\} \wedge T\{\sup_{s \in S_2} T'[\mu_{S_2}^*(s), \mu_{S_2}(s)], \mu_E(e)\}$$
$$= \underset{i=1}{\overset{2}{T''}} T\{\sup_{s \in S_i} T'[\mu_{s_i}^*(s), \mu_{S_i}(s)], \mu_E(e)\}. \qquad (7.22)$$

Recall that: T'' is a triangular norm which implements the **AND** connective (7.7), T' is the norm involved in the fuzzy inference (7.10) and T models the links as joint possibility distributions (7.7). Analogously to (7.13):

$$\alpha_j = \sup_s T'[\mu_j^*(s), \mu_j(s)], \quad j = 1, 2 \qquad (7.23)$$

$$\mu^*_{E_j}(e) = T\{\alpha_j, \mu_E(e)\} \tag{7.24}$$

$$\mu^*_E(e) = \underset{j=1}{\overset{2}{T''}}\mu^*_{E_j}(e) = \underset{j=1}{\overset{2}{T''}}[T\{\alpha_j, \mu_E(e)\}]. \tag{7.25}$$

B. Inference for node F.

$$\tilde{F}^*_k = [\tilde{E}^* \circ \tilde{R}^k_{E \times F}]U[\tilde{S}^*_3 \circ \tilde{R}^k_{S_3 \times F}], \quad k = 1 \div r \tag{7.26}$$

$$\mu^*_{F_k}(f) = \sup_{e \in E} T'\{\mu^*_E(e), \mu_R k(e, f)\}V \sup_{s \in S_3} T'\{\mu^*_{s_3}(s), \mu_{R_3 F} k(s, f)\}. \tag{7.27}$$

Each individual inference turns into:

$$\mu^*_F(f)$$
$$= \sup_e T'\{\mu^*_E(e), T[\mu_E(e), \mu_F(f)]\}V \sup_s T'\{\mu^*_{S_3}(s), T[\mu_{S_3}(s), \mu_F(f)]\}$$
$$= T[\sup_{e \in E} T'\{\mu^*_E(e), \mu_E(e)\}, \mu_F(f)]VT[\sup_{s \in S} T'\{\mu^*_{S_3}(s), \mu_{S_3}(s)\}, \mu_F(f)].$$
$$\tag{7.28}$$

C. Chaining.

By supplying in (7.28) μ^*_E from (7.25) the effect of *chaining* appears:

a)

$$\alpha_E$$
$$= \sup_{e \in E} T'\{\mu^*_E(e), \mu_E(e)\} = \sup_{e \in E} T'\{\underset{i=1}{\overset{2}{T''}}[T\{\alpha_i, \mu_E(e)\}], \mu_E(e)\}$$
$$= \sup_e T'\{T[T''(\alpha_1, \alpha_2), \mu_E(e)], \mu_E(e)\}. \tag{7.29}$$

α_3 is determined from (7.23) with $i = 3$. In fact, α_i $(i = 1 \div 3)$ are the *possibility measures of the detected symptoms* [22].

b)

$$\mu^*_F(f)$$
$$= T\{\alpha_E, \mu_F(f)\} \vee T\{\alpha_3, \mu_F(f)\}$$
$$= S''[T\{\alpha_E, \mu_F(f)\}, T\{\alpha_3, \mu_F(f)\}]. \tag{7.30}$$

In (7.30) by S'' the dual co-norm of the norm T'' is implementing the OR commective at the F nodes. Supplying in (7.30) α_E from (7.29):

$$\mu^*_F(f)$$
$$= S''[T\{\alpha_E, \mu_F(f)\}, T\{\alpha_3, \mu_F(f)\}] = S''T[\alpha_E, \alpha_3, \mu_F(f)]$$
$$= S''T\{\sup_e T'\{T[T''(\alpha_1, \alpha_2), \mu_E(e)], \mu_E(e)\}, \alpha_3, \mu_F(f)\}.$$
$$\tag{7.31}$$

The global fault possibility is:

$$\tilde{G}P_F = \bigcup_{l=1}^{r} F_k^*. \qquad (7.32)$$

Hints concerning a fault's size can be obtained by:

$$f = \text{DEF}(\tilde{G}P_F). \qquad (7.33)$$

7.4.2 Simplified Reasoning on the Tree

If T is lower semicontinuous (7.11) the supremum 'swallows' it in (7.31), together with $\mu_E(e)$ in (7.30) [23] as follows:

From (7.29)

$$
\begin{aligned}
\alpha_E^k \\
&= \sup_e T\{T[T''(\alpha_1^k, \alpha_2^K), \mu_{E_k}(e)], \mu_{E_k}(e)\} \\
&= T\{\sup_e T[\mu_{E_k}(e), \mu_{E_k}(e)], T''(\alpha_1^k, \alpha_2^K)\} T\{\sup_e \mu_{E_k}(e), T''(\alpha_1^k, \alpha_2^k)\} \\
&= T\{1, T''(\alpha_1^k, \alpha_2^k)\} = T''(\alpha_1^k, \alpha_2^k), \qquad (7.34)
\end{aligned}
$$

and from (7.30)

$$
\begin{aligned}
\mu_{F_k}^*(f) \\
&= S''[T\{\alpha_E^k, \mu_{F_k}(f)\}, T\{\alpha_3^k, \mu_{F_k}(f)\}] = S''T\{\alpha_E^k, \alpha_3^k, \mu_{F_k}(f)\} \\
&= S''T\{T''[\alpha_1^k, \alpha_2^k], \alpha_3^k, \mu_{F_k}(f)\} \\
&= T\{S''[T''(\alpha_1^k, \alpha_2^k), \alpha_3^k], \mu_{F_k}(f)\}. \qquad (7.35)
\end{aligned}
$$

It comes out that each k-explanatory possibility concerning a fault's existence $\tilde{F}_k^*$ (7.26) with the membership function $\mu_{F_k}^*(f)$ (7.27), is obtained by *weighting* (via the rule's association norm T) the *a priori* distribution of the fault variable $\tilde{F}_k$ with the aggregated measures α_j^k (7.23) of the detected symptoms (Figure 7.5).

The simplified reasoning procedure consists of the following steps:

a) Determine the k-possibility measures of the observed effects α_j^k (7.23);

b) Aggregate them on the tree's structure (7.35). The maximal aggregated value is PM:

$$
\begin{aligned}
PM \\
&= \sup_f \mu_{F_k}^*(f) = \sup_f T\{S''[T''(\alpha_1^k, \alpha_2^k,), \alpha_3^k], \mu_{F_k}(f)\} \\
&= T\{S''[T''(\alpha_1^k, \alpha_2^k), \alpha_3^k], \sup_f \mu_{F-k}(f)\} = T\{S''[T''(\alpha_1^k, \alpha_2^k), \alpha_3^k], 1\} \\
&= S''[T''(\alpha_1^k, \alpha_2^k), \alpha_3^k]. \qquad (7.36)
\end{aligned}
$$

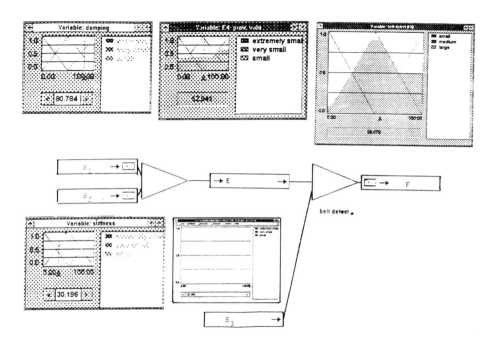

Figure 7.5: Simplified diagnostic reasoning using the FCM software package.

c) Weight $\tilde{F}_k$ (7.35) by the aggregated value for each k via the rule's norm T. Their union is GP (7.32).

d) DEF $(G\tilde{P})$ (7.33) gives a fault's state (f) most compatible with the explanation.

Figure 7.5 illustrates the implementation of the simplified reasoning strategy for the tree in Figure 7.2, by using the fuzzy software package Fuzzy Control Manager [4], for $T \equiv \min$ and $T''/S'' \equiv \min / \max$.

7.4.3   Significant Particular Cases

7.4.3.1

The *detected symptoms* s_j $(j = 1 \div 3)$ are *singletons*.

$$\mu_{s_j}(s_j) = 1, \mu_{s_j}(s) = 0, \qquad \text{for } s \neq s_j. \tag{7.37}$$

Then from (7.23) it comes out that

$$\alpha_j^k = \mu_{S_{jk}}(s_j), \tag{7.38}$$

i.e. the contribution of each symptom to the diagnostic reasoning equals the value of the possibility distribution $\tilde{S}_j$ in s_j. This is nothing else but the *possibility measure* of the detected symptom (Figure 7.5).

7.4.3.2

The event variable is a singleton

$$\mu_E(e') = 1, \mu_E(e) = 0 \qquad e \neq e' \tag{7.39}$$

$$
\begin{aligned}
\mu_F(f) \\
&= S''[T\{T'\{T[T''(\alpha_1, \alpha_2), 1], 1\}, \mu_F(f)\}, T\{\alpha_3, \mu_F(f)\}] \\
&= S''[T\{T''(\alpha_1, \alpha_2), \mu_F(f)\}, T\{\alpha_3, \mu_F(f)\}] \\
&= S''[T\{T''(\alpha_1, \alpha_2), \alpha_3, \mu_F(f)\}]. \tag{7.40}
\end{aligned}
$$

If $T \equiv$ MIN/product for each k-configuration, a fault's possibility results by aggregating on the tree's structure the α-cuts/products cutting/weighing $\tilde{F}$ at α_j^k $(i = 1 \div 3)$, *e.g.* in Figure 7.5, $k = 3$.

7.4.3.3

The *fault variable* is a *singleton*

$$\mu_F(f') = 1, \mu_F(f) = 0 \qquad f \neq f' \tag{7.41}$$

$$
\begin{aligned}
\mu_{F'}(f') \\
&= S''[T\{\sup_e T'\{T[T''(\alpha_1, \alpha_2), \mu_E(e)], \mu_E(e)\}, 1\}T\{\alpha_3, 1\}] \\
&= S''[\sup_e T'\{T[T''(\alpha_1, \alpha_2), \mu_E(e)], \mu_E(e)\}, \alpha_3\}]. \tag{7.42}
\end{aligned}
$$

If T, T' lower semicontinuous (satisfying (7.21)):

$$\mu_F^*(f') = S''[T''(\alpha_1, \alpha_2), \alpha_3] \tag{7.43}$$

$$PM = \mu_F^*(f) = S''[T''(\alpha_1, \alpha_2), \alpha_3]. \tag{7.44}$$

7.4.3.4

Both event and fault variables are singletons

$$
\begin{aligned}
\mu_{F'}(f') \\
&= S''[T\{T'\{T[T''(\alpha_1, \alpha_2), 1)], 1\}, 1\}, T\{\alpha_3, 1\}] \\
&= S''[T''(\alpha_1, \alpha_2), \alpha_3].
\end{aligned} \tag{7.45}
$$

The procedure reduces to a simple aggregation on the tree's structure of the possibility measures of the detected symptoms.

7.4.3.5

The above result is also obtained for the case where the fault and event variables are unknown, and accordingly modelled as particular fuzzy subsets which take a value of 1 everywhere in the universe.

7.5   Extension to Network Structures

7.5.1   Polytrees

Polytrees are the singly connected networks in which an effect has several possible causes. A general effect-node for polytrees is as shown in Figure 7.6. The k-joint possibility distribution of the effect:

$$
\Pi^k = \tilde{E}^k \times U[\bigcap_{j=1}^{p} \tilde{C}_j^k, \ldots, \bigcap_{j=q}^{m} \tilde{C}_j^k] \tag{7.46}
$$

corresponds to the rule:

$$
\text{IF } \tilde{E}^k \text{ THEN OR } \{\underset{j=1}{\overset{p}{\text{AND}}} \tilde{C}_j, \ldots, \underset{j=q}{\overset{m}{\text{AND}}} \tilde{C}_j\}. \tag{7.47}
$$

The *goal* of the inference can be expressed as "If $\tilde{E}^*$ is the possibility distribution of the observed effect, what can be said about the combination of possible causes?"

$$
\begin{aligned}
\Pi_C^k & \\
&= \tilde{E}^* \circ \Pi^k = \tilde{E}' \circ \left\{ \tilde{E}^k \times U\left[\bigcap_{j=1}^{p} \tilde{C}_j^k, \ldots, \bigcap_{j=q}^{m} \tilde{C}_j^k\right] \right\} \\
&= U\left[\bigcap_{j=1}^{p} \tilde{E}^* \circ (\tilde{E}^k \times \tilde{C}_j^k), \ldots, \bigcap_{j=q}^{m} \tilde{E}^*(\tilde{E}^k \times \tilde{C}_j^k)\right] \\
&= U\left[\bigcap_{j=1}^{p} \tilde{C}_j^{*k}, \ldots, \bigcap_{j=q}^{m} \tilde{C}_j^{*k}\right].
\end{aligned} \tag{7.48}
$$

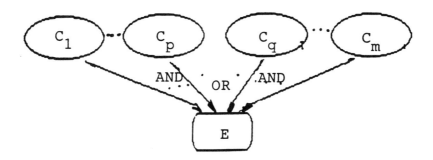

Figure 7.6: General node for polytrees.

The *decomposition theorem* [10, 25] reduces the inference to *local computations*:

$$\tilde{C}_j^{*k} = \tilde{E}^* \circ (\tilde{E}^k \times \tilde{C}_j^k) \tag{7.49}$$

as in the Bayesian inference, but, as opposite, these local computations are performed in parallel and propagated in the same direction (from the effects to the causes), and are finally re-composed into a *global* information Π_C^k via the nodes' connectives. The result of the inference is a *compound possibility distribution* in the product-universes of each individual cause:

$$\Pi_C^k = C_1 \times C_2 \times \cdots \times C_m \tag{7.50}$$

obtained by aggregating the results of all local computations on the node's structure. The individual possibility distributions for each cause node can be calculated by *projecting* the compound possibility distribution Π_C^k on each universe [6]:

$$\Pi_{C_1}^k = \underset{C_2,\ldots,C_m}{U} \Pi_C^k; \ldots; \Pi_{C_m}^k = \underset{c_1,\ldots,C_{m-1}}{U} \Pi_C^k. \tag{7.51}$$

The global possibilities for each cause and their sizes are determined according to (7.16) and (7.18).

Generalizing to polytrees, the *key idea* is to split the inference for each effect node into local computations according to the decompositional theorem and then to re-compose the local results by aggregation on the node's structure. The possibility distributions of the causes of interest are updated by *projecting* the compound possibility on the universe of the respective causes and propagated to the next causal level by repeating the same process. Con-

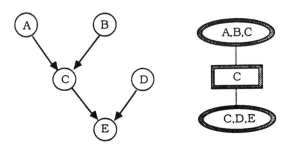

Figure 7.7: Polytree

sider the polytree in Figure 7.7.

$$
\begin{aligned}
\Pi_E^k \\
&= \tilde{E}^* \circ \tilde{E}^k \times (\tilde{C}^k U \tilde{D}^k) = [\tilde{E}^* \circ (\tilde{E}^k \times \tilde{C}^k)]U[\tilde{E}^* \circ (\tilde{E}^k \times \tilde{D}^k)] \\
&= \tilde{C}'^k U \tilde{D}'^k
\end{aligned}
\tag{7.52}
$$

$$
\tilde{C}^{*k} = \underset{D}{U}\Pi_E^k ; \tilde{D}^{*k} = \underset{C}{U}\Pi_E^k
\tag{7.53}
$$

$$
\Pi_C^k = \tilde{C}^{*k} \circ [\tilde{C} \times (\tilde{A} \cap \tilde{B})] = [\tilde{C}^{*k} \circ (\tilde{C} \times \tilde{A})] \cap [\tilde{C}^{*k} \circ (\tilde{C} \times \tilde{B})]
\tag{7.54}
$$

$$
\tilde{A}^{*k} = \underset{B}{U}\Pi_C^k ; \tilde{B}^{*k} = \underset{A}{U}\Pi_C^k .
\tag{7.55}
$$

As opposite to the trees, the inference is performed at each effect node via the key ideas: decomposition, re-composition and projection.

7.5.2   Multiply Connected Networks

The *decompositional inference theorem* [10] trivializes the inference in any kind of network. Consider for illustration the multiple-connected network in Figure 7.8. The network's k-possible configuration is globally modelled by the *compound possibility distribution*

$$
\Pi^k = \bigcap_{j=1}^{m} \tilde{E}_j^k \times \underset{i=1}{U} \tilde{C}_i ,
\tag{7.56}
$$

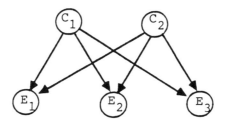

Figure 7.8: Multiply connected network.

which implements the *global rule*:

$$\text{IF } < E_1 \text{ AND } E_3 > \text{ THEN } < C_1 \text{ OR } C_2 > . \tag{7.57}$$

The inference flows in parallel through the network's links from the event nodes (E) to the explanatory nodes (C) aiming to compute their global k-compound possiblity

$$\Pi_{E_{1,2,3}} = \mathop{\circ}_{j=1}^{3} (E_j^*, \Pi^k). \tag{7.58}$$

The *decompositional inference theorem* splits the global inference in local inferences on each link:

$$\Pi_{E_{1,2,3}} = \mathop{u}_{i=1}^{2} \bigcap_{j=1}^{3} [\tilde{E}_j^* \circ (\tilde{E}_j^k \times \tilde{C}_i)]. \tag{7.59}$$

The *update possibility distributions* for each individual cause are obtained by projecting $\Pi_{E_{1,2,3}}$ on the universe of the respective cause:

$$C_1^{*k} = \mathop{U}_{C_2} \Pi_{E_{1,2,3}}; C_2^{*k} = \mathop{U}_{C_1} \Pi_{E_{1,2,3}}. \tag{7.60}$$

The inference process on the network in Figure 7.8 can be represented as a *multivariable fuzzy svstem*, Figure 7.9, with 3 inputs and 2 outputs [6].
Remark. In general the inference on any kind of possibilistic network can be represented as chained multivariable fuzzy systems, for each level of parent-child connections in one system.

7.5.3 Cyclic Networks

Modelling the links as non-directed joint possibility distributions and the global goal-oriented parallel inference approach for cause-possibility updating

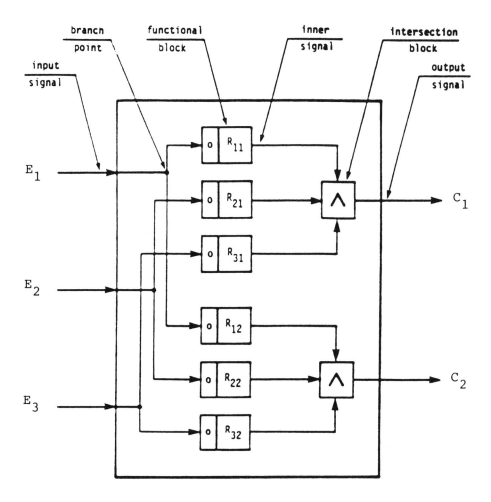

Figure 7.9: Multivariable fuzzy system corresponding to the network in Figure 7.8 (from [6]).

also solves in a very natural way the computational problems encompassed by the Bayesian inference in cyclic networks [14].

Consider the network in Figure 7.10. The Bayesian networks approach struggles to break the cycle by building an additional inference model, called a *junction* tree [11], on which a tree-like inference is then to be performed. However, this is not always easy to do, and usually requires additional so-called *fill-ins* [8], *i.e.* supplementary links to be added to the initial network. Consider that the observable effect is C_3 in Figure 7.10, in the form of the new evidence $\tilde{C}_3^*$, and the goal is to determine the possibility for cause C_1 under these circumstances. The procedure follows two different courses according to the dual role of C_3, which is at the same time cause and effect of C_1. On one

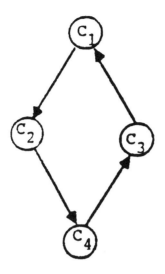

Figure 7.10: Cyclic network.

side $\tilde{C}_3^*$ modifies the possibility distribution $\tilde{C}_1$ as its *cause*:

$$\tilde{C}_{1(3)}^{*k} = \tilde{C}_3^* \circ \Pi_{31}^* = \tilde{C}_3^* \circ (\tilde{C}_3^k \times \tilde{C}_1^k). \qquad (7.61)$$

This modified evidence will update the joint possibility modelling the link $C_1 - C_2$:

$$\Pi_{12}^k = \tilde{C}_2^k \times \tilde{C}_{1(3)}^{*k}. \qquad (7.62)$$

On the other side, $\tilde{C}_3^*$ is *explained* by C_1 as its *effect*:

$$\tilde{C}_4^{*k} = \tilde{C}_3^* \circ \Pi_{34}^k = \tilde{C}_3^* \circ (\tilde{C}_3^k \times \tilde{C}_4^k) \qquad (7.63)$$

$$\tilde{C}_2^{*k} = \tilde{C}_4^* \circ \Pi_{42}^k = \tilde{C}_4^* \circ (\tilde{C}_4^k \times \tilde{C}_2^k) \qquad (7.64)$$

$$\tilde{C}_1^{*k} = \tilde{C}_2^{*k} \circ \Pi_{12}^k = \tilde{C}_2^{*k} \circ (\tilde{C}_2^k \times \tilde{C}_{1/3}^{*k}). \qquad (7.65)$$

7.6 Illustration and Investigations on a Tutorial Example

7.6.1 The Diagnostic Model

Consider now the simple example in Figure 7.2, which is an extract from a more complex fault tree designed for a feed drive [7]. The membership

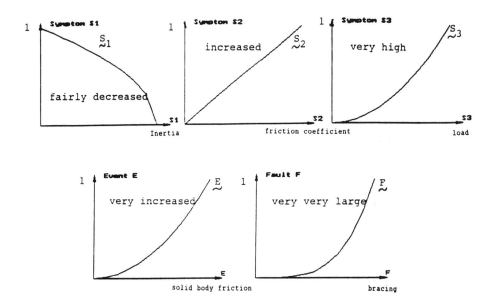

NODE	UNIVERSE	FUZZY SUBSET/LINGUISTIC VAR.
F	bracing of the slideways	F ≡ *very very large*
E	solid body friction	E ≡ *very increased*
S_1	rotational inertia	S_1 ≡ *fairly decreased*
S_2	friction coefficient	S_2 ≡ *increased*
S_3	load	S_3 ≡ *very high*

Figure 7.11: The fuzzy subsets assigned to the nodes in Figure 7.2.

functions assigned to the nodes of the possibilistic tree as well as the linguistic description of the respective fuzzy subsets are illustrated in Figure 7.11.

In the following the calculations will be made on discrete normalized universes with a step of 0.1 (*i.e.* the values of the universes will be from 0 to 1 with 10 points of discretization: 0,1 0,2 0,3...0,9, 1). For an optimal visualization of both final and intermediate results the shapes of the membership functions were chosen such that they cover the whole universes of discourse. In particular, the fuzzy subset $\tilde{E}$ (assigned to the event node E) representing the linguistic variable 'very increased' was defined by using Zadeh's definition of the linguistic hedge 'very' as a nonlinear operator which modifies the meaning of its operand (in our case the linguistic variable 'increased') in a specified fashion [30].

By assigning to each node of the tree in Figure 7.2 the corresponding fuzzy subset, the links will bear the possibilistic meaning of the corresponding fuzzy IF-THEN rules:

IF <rotational inertia is *decreased* > THEN <solid body friction is *very increased*>;

IF <friction coefficient is *increased*> THEN <solid body friction is *very increased*>;

IF <solid body friction is *very increased*> THEN <bracing of the slideways are *very very large*>;

IF < load is *high* > THEN <bracing of the slideways are *very very large*>.

Regarded as joint possibility distributions on the product universes of the antecedent and the concequent, the above fuzzy IF-THEN rules will induce the fuzzy relations:

$$\tilde{R}_{S_1 \times E} : S_1 \times E \to [0, 1]; \tag{7.66}$$

$$\tilde{R}_{S_2 \times E} : S_2 \times E \to [0, 1]; \tag{7.67}$$

$$\tilde{R}_{E \times F} : E \times F \to [0, 1]; \tag{7.68}$$

$$\tilde{R}_{S_3 \times F} : S_3 \times F \to [0, 1]; \tag{7.69}$$

with the joint possibility distribution functions defined as discrete two-valued membership functions:

$$\mu_{R_{S_1 \times E}}(s, e) = \min\{\mu_{S_1}(s), \mu_E(e)\}, \quad s \in S_1 \text{ and } e \in E; \tag{7.70}$$

$$\mu_{R_{S_2 \times E}}(s, e) = \min\{\mu_{S_2}(s), \mu_E(e)\}, \quad s \in S_2 \text{ and } e \in E; \tag{7.71}$$

$$\mu_{R_{E \times F}}(e, f) = \min\{\mu_E(e), \mu_F(f)\}, \quad e \in E \text{ and } f \in F; \tag{7.72}$$

$$\mu_{R_{S_3 \times F}}(s, f) = \min\{\mu_{S_3}(s), \mu_F(f)\}, \quad s \in S_3 \text{ and } f \in F; \tag{7.73}$$

presented in Figure 7.12.

To build the possibilistic diagnostic model (which will consist in this simplified case of a single row, mapping the symptoms S_1, S_2 and S_3 to the fault F, Figure 7.13) the part of the tree containing the chains $S_i \to E \to F, i = 1, 2$ has to be compressed by fuzzy composition to the direct symptoms-fault fuzzy relations:

$$\tilde{R}_{S_1 \times F} = \tilde{R}_{S_1 \times E} \circ \tilde{R}_{E \times F} \tag{7.74}$$

and

$$\tilde{R}_{S_2 \times F} = \tilde{R}_{S_2 \times E} \circ \tilde{R}_{E \times F} \tag{7.75}$$

with the membership functions:

$$\mu_{R_{S_1 \times F}}(s, f)$$
$$= \max_e \{\min[\mu_{R_{S_1 \times E}}(s, e), \mu_{R_{E \times F}}(e, f)]\} \quad \text{where } s \in S_1, e \in E, f \in F \tag{7.76}$$

$$\mu_{R_{S_2 \times F}}(s, f)$$
$$= \max_e \{\min[\mu_{R_{S_2 \times E}}(s, e), \mu_{R_{E \times F}}(e, f)]\} \quad \text{where } s \in S_2, e \in E, f \in F \tag{7.77}$$

presented in Figure 7.14 as discrete two-valued membership functions.

7.6.2   Approximate Reasoning on the Possibilistic Tree

For an elloquent illustration of the process of diagnostic reasoning the detected values of the symptoms S_1, S_2 and S_3 were chosen to be such that they match the fuzzy subsets $\tilde{S}_1$, $\tilde{S}_2$ and $\tilde{S}_3$, respectively: $\tilde{S}_1^*$ with a degree $\alpha_1 = 0.84$, $\tilde{S}_2^*$ with $\alpha_2 = 0.5$ and $\tilde{S}_3^*$ with $\alpha_3 = 0.36$ (Figure 7.15).

By approximate reasoning on the possibilistic diagnostic model (Figure 7.16), with the detected symptoms, local explanatory possibility distributions of the fault by symptoms S_1, S_2 and S_3 are

$$\tilde{F}_{S_j} = \tilde{S}_j^* \circ \tilde{R}_{S_j \times F}; \quad j = 1 \div 3 \tag{7.78}$$

by which each detected symptom brings information concerning a fault's F existence are obtained.

The corresponding possibility distribution functions:

$$\mu_{F_{S_j}}$$
$$= \max_s \min\{\mu_{S_j}(s), \mu_{R_{S_j \times F}}(s, f)\}, \quad s \in S_j \text{ and } f \in F; \quad j = 1 \div 3 \tag{7.79}$$

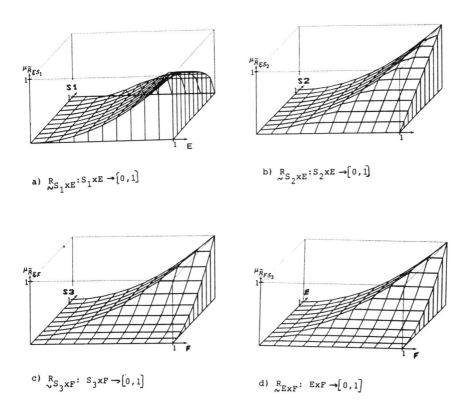

a) $\underset{\sim}{R}_{S_1 \times E} : S_1 \times E \rightarrow [0,1]$

b) $\underset{\sim}{R}_{S_2 \times E} : S_2 \times E \rightarrow [0,1]$

c) $\underset{\sim}{R}_{S_3 \times F} : S_3 \times F \rightarrow [0,1]$

d) $\underset{\sim}{R}_{E \times F} : E \times F \rightarrow [0,1]$

Figure 7.12: The fuzzy relations modelling the links in Figure 7.2.

are illustrated in Figure 7.16 attached to the bottom nodes. Figure 7.17 illustrates the possibility distribution $\tilde{F}_{S_2}$ computed for $\tilde{S}_2^*$ matching the rule $\tilde{R}_{S_2 \times F}$ with degrees ranging from 0 to 1 with a step of 0,1. It is easy to notice that as the degree of matching increases, the importance of the information by which S_2 contributes (quantified by the membership function) increases as well. This is the first step towards *managing the dynamics of the fuzzy information* (fault's dynamics), and will be further exploited in the next subsection in forecasting the fault tendency, as well as in coping with other complex diagnostic tasks.

	S_1	S_2	S_3
F	$\underset{\sim}{R}_{S_1 \times F}$	$\underset{\sim}{R}_{S_2 \times F}$	$\underset{\sim}{R}_{S_3 \times F}$

Figure 7.13: The possibilistic model for the example in Figure 7.2.

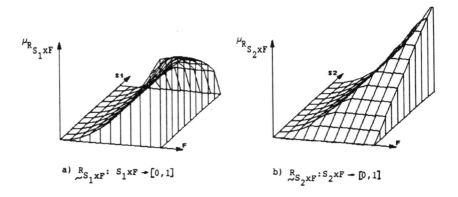

a) $\underset{\sim}{R}_{S_1 \times F}$: $S_1 \times F \rightarrow [0,1]$ b) $\underset{\sim}{R}_{S_2 \times F}$: $S_2 \times F \rightarrow [0,1]$

Figure 7.14: The compressed fuzzy relations.

7.6.3   Aggregation on the Tree's Structure

The global fault possibility $G\tilde{P}$ is computed by aggregating the local possibility distributions according to the formula:

$$F = \text{OR}\{\text{AND}[S_1, S_2].S_3\} \tag{7.80}$$

extracted from the tree's structure. The distribution function of the global fault possibility is presented in Figure 7.16, attached to the top node.

The possibility measure of fault F for this example results to be:

$$PM_F = \max_f \mu_{GP}(f) = 0.5. \tag{7.81}$$

Figure 7.18 illustrates the global explanatory possibility GP_F when symptom S_2^* ranges, with the local possibility as in Figure 7.17. It is easy to notice that the more the degrees by which S_2 matches increase (*i.e.* the more S_2 is

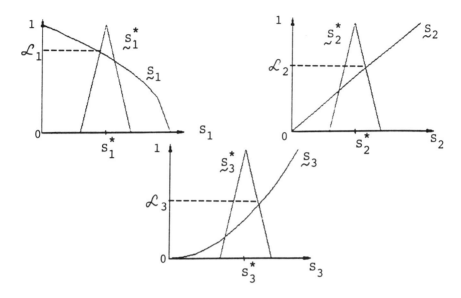

Figure 7.15: The inputs (S_j^*) matching the model(S_j).

detected), the more the global fault possibility increases (*i.e.* the more the respective symptom contributes to the fault's identificability). This will be further exploited in subsection 7.6.6 in managing a fault's dynamics.

7.6.4 Validation

Consider now that fault F is a candidate. To validate fault F the global fault possibility in $\tilde{G}P_F$ (attached to the top node in Figure 7.16) has to be projected by generalized *modus tollens* on each of the fuzzy relations in Figure 7.12 [13] representing the elements of the possibilistic diagnostic model. The projections:

$$\tilde{S}'_j = \tilde{R}_{S_j \times F} \circ \tilde{G}P; \quad i = 1 \div 3 \qquad (7.82)$$

have the *possibilty distribution functions*:

$$\mu'_{S_j}(s)$$
$$= \max_f \min\{\mu_{R_{S_j \times F}}(s, f), \mu_{GP}(f)\}, \quad s \in S_j, f \in F; \quad j = 1 \div 3, \qquad (7.83)$$

and are presented in Figure 7.19 as α-cuts, cutting the initial data $\tilde{S}_1$, $\tilde{S}_2$ and $\tilde{S}_3$ at the aggregated possibility measures of the detected symptoms

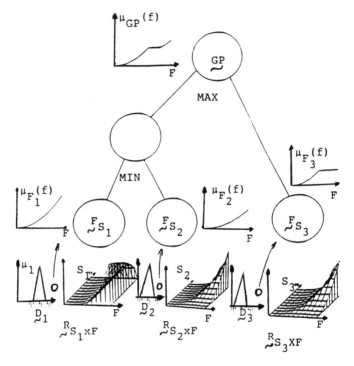

Figure 7.16: Diagnostic reasoning on the fuzzy fault tree.

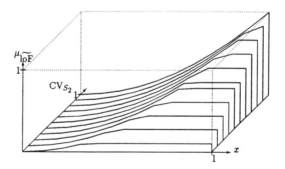

Figure 7.17: Possibility distribution of fault F by symptom S_2.

S_1^*, S_2^* and S_3^*. To prove this, let us supply in (7.35) from subsection 7.4.2: $S''/T'' \hat{=} \max / \min; T \hat{=} \min; F_k^* \hat{=} GP_F; F_k \hat{=} F; \alpha_j^k \hat{=} \alpha_j$, with $\alpha_{1,2,3}$ defined as in (7.13), subsection 7.3.1, and illustrated in Figure 7.15 for the particular

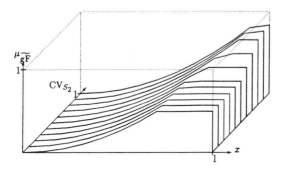

Figure 7.18: Global possibility of F when S_2 ranges.

implementation when the degrees of matching data-model are: 0.84 for the detected symptom S_1; 0.5 for S_2 and 0.36 for S_3:

$$\mu_{GP}(f) = \min\{\max[\min(\alpha_1, \alpha_2), \alpha_3], \mu_F(f)\} \qquad (7.84)$$

Denoting by α the aggregated possibility measures α_j, $j = 1 \div 3$:

$$\alpha = \max[\min(\alpha_1, \alpha_2), \alpha_3] \qquad (7.85)$$

(7.83) becomes:

$$
\begin{aligned}
\mu'_{S_j}(s) \\
&= \max_f \min\{\min[\mu_{S_j}(s), \mu_F(f)], \min[\alpha, \mu_F(f)]\} \\
&= \min\{\mu_{S_j}(s), \max_f \min[\alpha, \mu_F(f)]\} \qquad (7.86)
\end{aligned}
$$

q.e.d.

To calculate the *distance* between these projections α-cuts and the initial diagnostic knowledge, represented by the fuzzy subsets $\tilde{S}_1$, $\tilde{S}_2$, $\tilde{S}_3$ one can choose for example the generalised Hamming distance ([9], p.19):

$$H_{S_1} = \frac{1}{10} \sum_{\substack{s=1 \\ (s \in S_1)}}^{10} (|\mu_{S_1}(s) - \mu'_{S_1}(s)|) = 0.213 \qquad (7.87)$$

$$H_{S_2} = \frac{1}{10} \sum_{\substack{s=1 \\ (s \in S_2)}}^{10} (|\mu_{S_2}(s) - \mu'_{S_2}(s)|) = 0.136 \qquad (7.88)$$

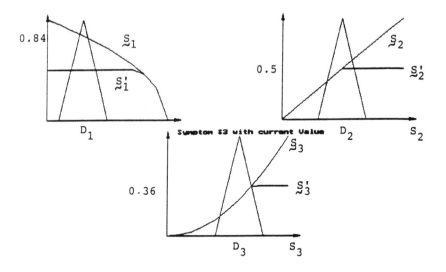

Figure 7.19: The projections as alpha-cuts of the initial data.

$$H_{S_3} = \frac{1}{10} \sum_{\substack{s=1 \\ (s \in S_3)}}^{10} (|\mu_{S_3}(s) - \mu'_{S_3}(s)|) = 0.086. \qquad (7.89)$$

As in our example symptoms, S_2 and S_3 are matching with non-validable values (respectively with 0.5 and 0.36, Figure 7.19) the fault is not validated, and this is reflected by the fact that the contribution of symptom S_1 is *minimal*, so the respective distance is *maximal*. On the other hand, the contribution of S_3 is maximal to the non-validation of the global fault, so its distance is minimal. This may at a first glance seem paradoxical, but in fact it is a natural concequence, as long as the symptom S_1 could have a maximal contribution to the validation, therefore naturally having a minimal contribution to the *non*-validation. On the contrary, the contribution of symptom S_3 to the non-validation is maximal and concequently the respective distance H_{S_3} is minimal.

In the example considered, S_2, which matches with a degree of 0,5, keeps the balance between the high degree by which S_1 matches (0.87) and the very low one of S_3 (0.36). Therefore the detection of S_2 appears to be crucial for the fault detection (validation) as it results as well from Figure 7.20a), b), c) which illustrates $\tilde{S}'_1$, $\tilde{S}'_2$, $\tilde{S}'_3$ in case that the degree of matching of symptom S_2, denoted by D_2, ranges from 0 to 1 (according to Figure 7.18).

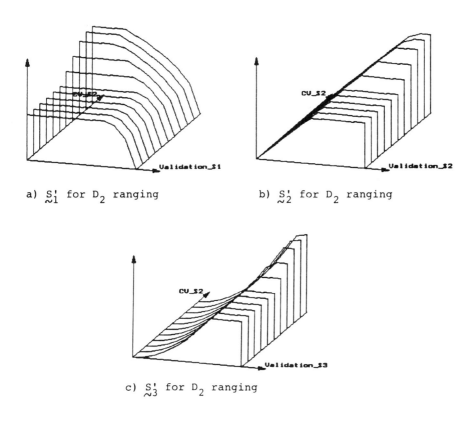

a) $\underset{\sim}{S}'_1$ for D_2 ranging

b) $\underset{\sim}{S}'_2$ for D_2 ranging

c) $\underset{\sim}{S}'_3$ for D_2 ranging

Figure 7.20: The projections of $G\tilde{P}$ on the possibilistic model for D_2 (the detected value of symptom S_2) ranging from 0 to 1.

7.6.5   Implementation Considerations

In addition to its efficiency, the simplicity of the proposed fault decision approach is well suited for engineering applications. It has been successfully tested on simulated processes (a feed drive and a DC motor [29, 27]).

For the implementation a $C++$ *object-oriented interface* (Figure 7.21) between the software packages: MATLAB — with its simulation facility SIMULINK (which models the correct system's behaviour to perform the *discrepancy detection* for symptoms generation), and Fuzzy Control Manager (FCM - TRANSFERTECH Germany [4], which implements the fuzzy diagnostic strategy), has been designed.

Figures 7.22 a) and b) detail the implementation of the diagnostic strategy for the example in Figure 7.2. To each cause node an inference block (symbolized by a triangle in part (a) of the figure) is allocated. To avoid the

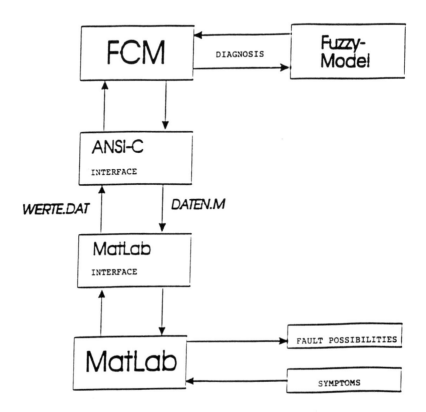

Figure 7.21: Software implementation of the diagnostic system.

effects of the implicit defuzzification mechanism embedded within the fuzzy inference blocks (because the fuzzy software is dedicated mainly to *control* applications), a trick has been made by putting the cause variable at the input of the inference block as well. In this way the inferred possibility distribution of the chaining variable (*e.g.* $\tilde{E}'$) can be kept (in addition to its defuzzified value) and propagated through chaining as input to the next inference block.

To benefit from the defuzzification facilities embedded within the fuzzy software package FCM, a supplementary trick has been used for the implementation: instead of defining, for example, 17 output fuzzy variables in case of the DC motor's diagnosis (as 17 faults have been considered for the test process [29]), only one global output variable has been designed, by having 17 linguistic terms, defined as *singletons*. In this way, by simply choosing as the defuzzification option the *maximum height*, the linguistic term with the maximal global possibility is automatically selected, and the corresponding fault identified.

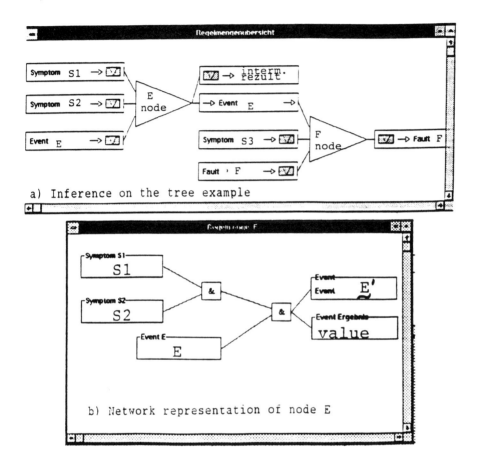

Figure 7.22: Implementation details for the example in Figure 7.2.

The system works on any kind of PC-AT compatible with Windows 3.1. All the faults artificially injected as parameter changes within the test processes have been correctely diagnosed.

7.6.6 Managing a Fault's Dynamics

7.6.6.1 Dynamic Fuzzy Fault Tree

Consider now the dynamic evolution of the detected symptoms, $s_j(t)$ ($j = 1 \div 3$) in time (Figure 7.23) as a reflection of the fast development of the *incipient fault F*. By *matching* $s_j(t)$ with the multidimensional-multivariable *approximate reasoning model* induced by the fuzzified fault tree (recall section 7.2) (Figure 7.24), the possibility distributions $\tilde{F}_{S_j}$ (7.78) by which the

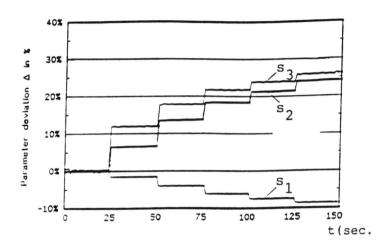

Figure 7.23: Dynamic evolution of the symptoms.

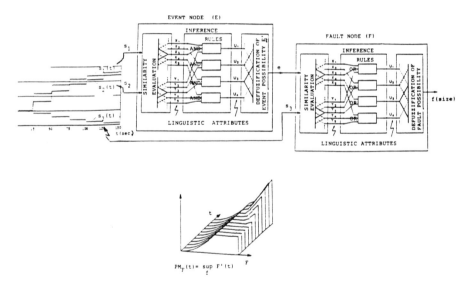

Figure 7.24: Representation of the multidimensional-multivariable approximate reasoning model corresponding to Figure 7.2.

evolution in time of the symptoms brings information concerning a fault's gradual development, are determined, in Figure 7.25 attached to the bottom nodes. The *fault's dynamics* will be reflected through the evolution in time of its global possibility distribution $GP(t)$ (illustrated in Figure 7.25 at the top node) obtained by aggregation of $\tilde{F}_{S_i}(t)$ on the tree (7.80). In this way, a *dynamic fuzzy fault tree*, which models the fault dynamics via processing of

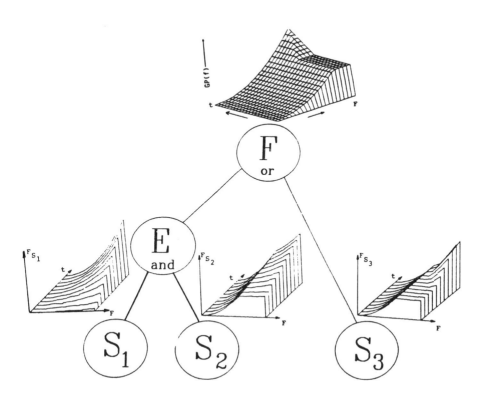

Figure 7.25: Dynamic fuzzy fault tree.

the dynamic fuzzy information on the structure of the crisp fault tree, naturally emerges from this multivariable-multidimensional approximate reasoning diagnostic strategy.

7.6.6.2   Incipient Fault Dynamics

Figure 7.26 illustrates the dynamics of fault F in our example obtained by making the aggregation via the $./\oplus$ fuzzy connectives product/bounded sum, for the same evolutions of the detected dynamic symptoms $s_i(t)$ ($i = 1 \div 3$) (Figure 7.23). By comparison with the $GP(t)$ in Figure 7.25 obtained via the MIN/MAX connectives, it is clear that the *possibilistic* MIN/MAX pair, due to its high nonlinearity, makes a sharper discrimination, by being more adequate for *incipient faults detection*.

7.6.6.3   Monitoring Intermittent Faults

The fault dynamics when the intermittent symptom $s_1(t)$ (Figure 7.27) was detected is presented in Figure 7.28a) under the MIN/MAX connectives and

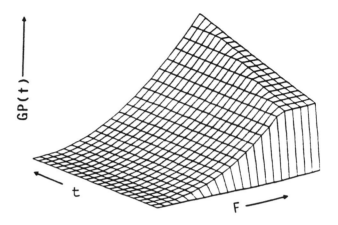

Figure 7.26: Dynamics of the fault's possibility by product/bounded sum.

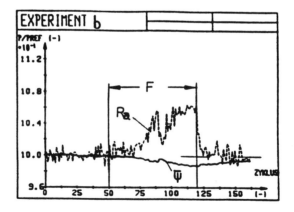

Figure 7.27: Intermittent symptom.

b) under the product/bounded sum (probabilistic pair). (The other two symptoms are constant.) The *probabilistic ./+* pair, due to its effect of *interactivity*, allowing *all* the symptoms to contribute with information in monitoring the dynamics, is more adequate for *diagnosing intermittent faults.*

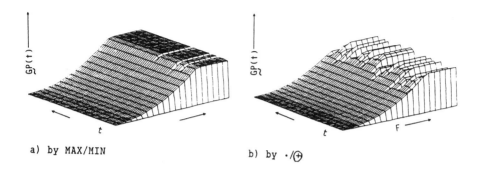

a) by MAX/MIN

b) by $\cdot/\oplus$

Figure 7.28: Monitoring intermittent faults

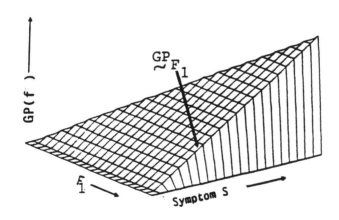

Figure 7.29: Global possibility of F_1 depending on symptom S.

7.6.6.4   Multiple faults with compensatory effects

Consider for illustration two faults of a feed drive of a machine-tool which show up via *opposite symptoms* S and $\bar{S}$ [7]:

- F1: 'Lack of lubrication' $\rightarrow S$ – 'increase of coulomb friction coefficient' (Figure 7.29);

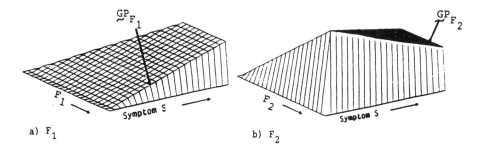

Figure 7.30: Influence of the compensatory effect decreasing detectability.

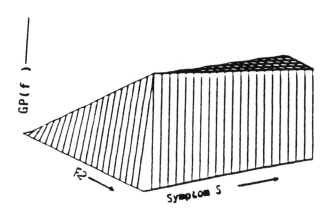

Figure 7.31: Improred detection for F_2.

- F2: 'defect of the belt' → $\bar{S}$ – 'decrease of coulomb friction coefficient'.

Figures 7.30a) and b) illustrate, via *fuzzy information*, how the detectability of the faults decreases due to the compensation of the opposite symptoms S and $\bar{S}$, when both faults occur simultaneously. By acquiring supplementary information from the other symptoms a considerable improvement in the detection of fault F_2 is achieved (Figure 7.31), while in case of F_1 no improvement is noticed, as (according to the causal structure which determines the aggregation formula) symptom S brings the dominant information in this case.

7.7   Conclusions

Within an attempt to emulate the mechanisms involved in a human diagnostician's judgement when handling the complexity of the diagnostic phenomenology in attempting to solve the diagnostic problem, an approach for *diagnostic reasoning on possibilistic networks* has been proposed. It can be used in the design of approximate reasoning-based diagnostic systems.

It was shown that possibilistic networks are multivariable fuzzy systems that constitute fair representations of the soft connectionistic mental models which support a diagnostician's reasoning. On a global scale, the diagnosis is made by approximate matching of the information structured in the soft network with the symptoms detected. This in turn emulates, in a fair way, the global analogical reasoning mechanism based on similarity evaluation, through which the diagnostician finds the explanation by evaluating the extent to which the available data (symptoms) fit his/her current mental model.

The explanation is found by parallel activation of several possible configurations of diagnostic knowledge and implicit selection of the most compatible (similar) solution with the observations. The information flows from the symptoms to their explanation within a goal-oriented strategy, using the network's structure as a relaxed and flexible guideline, adaptive to the *global* goal (establishment of the explanatory possibility of the diagnosis.) This fairly emulates the way in which humans deal with the complexity of the diagnostic phenomenology.

The following characteristics of the inference on the possibilistic networks have been underlined:

- *Diagnostic reasoning mechanism.* The explanatory possibility distribution of the diagnosis is globally calibrated to the information brought by all its observed effects, by local parallel evaluation of their similarity to the prior correlated information encoded on the links. Several possible configurations are considered in parallel, and the relevant information is extracted within an implicit act of interpolative reasoning. This mechanism follows human explanatory judgement which is usually made by a *global* rather than an analytical assessment [1]. Humans succeed in making a quick assessment of an explanatory situation pattern, followed by an immediate categorization of the situation with finding an associative explanation, by analogy with what they know. (The knowledge being encoded within the soft network.)

- *Ultimate goal of the reasoning.* Calibration of the possibilistic network to the global configuration most compatible with the observed effects, via parallel activation of all possible configurations (rules) by all the observed effects. This emulates the transformation of the available data (symptoms) to fit to the available mental model (here regarded as an act of implicit adaptiveness of the network's information to the observed symptoms/data). The calibration is implicitly embedded within the absorption (7.14) of the new evidence (7.13) on the (fuzzy) subset of the explanation.

- *Key Issue.* To consider the whole network as a guideline for the explanatory flow of the inference from the symptoms to their diagnosis. This follows the goal-orientation of the human explanatory judgement which is unconsciously driven by the desired result [1].

The decompositional inference theorem trivializes the inference in any kind of network via parallel local inferences on each link followed by aggregations on the logical structure of the nodes and projections on the universe of the explanatory variable. This turns into a considerable reduction of the computational complexity.

The ability to calibrate ON-LINE the structure of the diagnostic model [16], making it adaptive to dynamic changes in the behaviour of the diagnosed system, from the perspective of designing and implementing a reconfigurable diagnostic strategy [28].

Based on these theoretical results, a dynamic-possibilistic approach for fault trees processing in diagnostic decision making has been proposed. Able to handle numerically a huge amount of vague information by extracting and propagating only its relevant part, the proposed multidimensional-multivariable approximate reasoning strategy for diagnostic decision making is a powerful tool for managing linguistic information within diagnostic expert systems. By quantifying the universe of the variables, hints concerning a fault's size can be extracted directly from the reasoning procedure (by defuzzification).

In addition to its efficiency, the simplicity of the proposed methodology (which is developed by using mostly primitive but application-effective principles of fuzzy sets theory and its derivate reasoning system — fuzzy logic) makes it very well suited for practical (engineering as well as medical) applications. The importance of fuzzy information processing based on the novel concept of dynamic fuzzy fault tree for managing the complexity involved in diagnosing incipient/intermittent/multiple faults has been illustrated by examples.

References

[1] AGARD (1995) Advisory Group for Aerospace Research & Development, NATO Report AGARD-AR-325 7 Rue Ancelle, 92200 Neuilly-sur-Seine, France.

[2] Bezdek J. (1993) 'Fuzzy Models: What are they and why', Editorial, *IEEE Transactions of Fuzzy Systems*, 1 (1).

[3] Dubois D. and Prade H. (1988) *Possibility Theory*, Plenum Press, NY.

[4] FCM (1995) *Fuzzy Control Manager - User's Manual, V. 4.1.*, TRANS-FERTECH GmbH, Cyriaksring 9A, 38118 - Braunschweig, Germany.

[5] Gottwald S. (1993) *Fuzzy Sets and Fuzzy Logic*, Vieweg Verlag, Wiesbaden.

[6] Gupta N.N., Kiszka J.B. and Trojan G.M. (1986) Multivariable structure of fuzzy control systems. *IEEE/SMC*, 16: 638–656.

[7] Isermann R. (1993) Fault diagnosis by parameter estimation and knowledge processing. *Automatica*, 29 (4).

[8] Jensen F.V. (1993) *Introducton to Bayesian Networks*, HUGIN, Denmark.

[9] Kaufmann A. (1975) *Introduction to the Theory of Fuzzy Subsets - Vol.1*, Academic Press, NY, 1–38, 46–69.

[10] Kosko B. (1992) *Neural Networks and Fuzzy Systems*, Prentice Hall.

[11] Lauritzen S.L. and Spiegelhalter D.S. (1988) Local computations with probabilities on graphical structures and their application to expert systems. *J. P. Statistical Society*, B 50: 157–224.

[12] Leitch R.R. and Quek H.C. (1991) A behaviour classification for integrated process supervision. In *Proceedings of IEE International Conference on Control* (2) 127–133.

[13] Linkens D.A. and Nie J. (1992) A unified real time approximate reasoning approach for use in intelligent control. *Int. J. of Control*, (56) 2: 359–361.

[14] Pearl J. (1988) *Probabilistic Reasoning in Intelligent Systems – Networks of plausible inference*, Morgan Kaufmann.

[15] Ruspini E.H. (1991) Approximate Reasoning: Past, Present, Future. *Journal of Information Sciences* 57–58: 297–317.

[16] Smith M.H. (1995) Tuning fuzzy aggregation operators using fuzzy metarules. In *Proceedings of the IEEE/SMC'95 Conference "Intelligent Systems of the 21st Century"*, 22–25 October, Vancouver, Canada.

[17] Ulieru M. (1993) A Fuzzy Logic Based Computer Assisted Fault Daignosis System. In *Proceedings of TOOLDIAG'93 International Conference on Fault Diagnosis*, April 5–7, Toulouse, France, 689–699.

[18] Ulieru M. (1993) Processing fault trees by approximate reasoning. In *Proceedings of the 12th IFAC World Congress*, (9) 221–224, Sydney, Australia. Pergamon Press.

[19] Ulieru M. (1993) From fault trees to fuzzy relations in managing heuristics for technical diagnosis. In *Proc. IEEE/SMC'93 Conf.*, (1) 743–748, Le Touquet, France.

[20] Ulieru M. (1993) A Unified Fuzzy Approach for Diagnostic Decision by Causal Networks. In *Proc. QUARDET'93*, Barcelona, June 16–18, 672–682.

[21] Ulieru M. (1993) From Boolean Probabilistic to Fuzzy Possibilistic Fault Tree Processing in Diagnostic Decision Making. In *Proceedings of EU-FIT'93*, First European Conaress on Fuzzy and Intelligent Technologies, (1) 408–414, Aachen, Germany, September 7–10.

[22] Ulieru M. (1994) Approximate reasoning approaches for diagnostic decision - Vis-a-vis Probabilistic and Possibilistic techniques. In *Proceedings of SAFEPROCESS'94*, Espoo, Finland, June 13–15 Pergamon.

[23] Ulieru M. (1994) Diagnosis of Continuous Dynamic Systems Based on Fuzzy Information Processing. In *Proceedings EUFIT'94*, September 21–23, Aachen.

[24] Ulieru M. (1994) Diagnosis by Approximate Reasoning on Dynamic Fuzzy Fault Trees. In *Proceedings Fuzz-IEEE'94*, Orlando, June 26–29.

[25] Ulieru M. (1994) Fuzzy Logic in Diagnostic Decision - Possibilistic Networks. In *Proceedings IPMU'94*, July, Paris, (1) 96–101.

[26] Ulieru M. (1994) A Fuzzy Logic Based Computer Assisted Fault Diagnosis System. *Revue Europeene Diagnostique et Surrete du Fonctionnement*, (3) 3: 405–441.

[27] Ulieru M. (1995) Fuzzy Logic in Diagnostic Decision: Possibilistic Networks. Dissertation, Technische Hochschule Darmstadt.

[28] Ulieru M. and Smith M. (1996) A Fuzzy Approach for the Design of Reconfigurable Diagnostic Systems. In *Proceedings of NAFIPS'96*, June, Berkeley, CA.

[29] Ulieru M. and Isermann R.(1993) Design of a fuzzy logic -based diagnostic model for technical processes. *Fuzzy Sets & Systems* 58: 249–271.

[30] Zadeh L.A. (1975) The concept of a linguistic variable and its application to approximate reasoning - Parts I - III. *Information Sciences*, 8 and 9: 199–249, 301–357, 43–80.

[31] Zadeh L.A. (1978) Fuzzy Sets as a Basis for a Theory of Possibility. *Int. Journal of Fuzzy Sets and Systems*, 1: 3–28.

[32] Zimmermann H.J. (1991) *Fuzzy Set Theory and its Applications, Second Edition*, Kluwer Academic, 57–67, 91–105.

[33] Zimmermann H.J. and Zysno P. (1983) Decisions and evaluations by hierarchical aggregation of information. *International Journal of Fuzzy Sets and Systems*, (10) 243–260.

8

Fuzzy Linguistic Reasoning and Sentence Interpretation

J.F. Baldwin, B.W. Pilsworth

8.1 Introduction

The expressiveness and elasticity of natural language utterances comes from
the fusion of syntactic structure, semantic content and a use of words and
phrases which is rich in metaphors, analogies and descriptions which are im-
bued with ambiguity, imprecision, vagueness and other forms of uncertainty.
Most conventional artificial intelligence treatments of natural language modelling focus on the structural and semantic aspects using representations such
as first order logic, semantic nets, frames, scripts and conceptual graphs.
However, language can be both expressive and efficient in its representation
of knowledge by using uncertain forms to provide summarising descriptions
which cover a broad span of relevant knowledge concisely and intelligibly.
Fuzzy, probabilistic and evidential reasoning are necessary to model the various forms of uncertainty in natural language statements, so that these can be
interpreted appropriately when used in such applications as browsers in fuzzy
databases, look-ups for intelligent manuals and front-end interfaces to expert
systems.

The language Fril: A Support Logic Programming System [1]-[4], provides
a powerful and versatile tool for modelling: probabilistic uncertainty, through
its support pair representation and support logic inference mechanism, and
fuzzy concepts through its representations of fuzzy sets and partial matching by semantic unification. An important component in the modelling of
linguistic statements with uncertainty is the recognition of the constituents

and scope of the sources of uncertainty. This involves structural analyses of sentences using both syntactic parses and semantic interpretations using knowledge of word meanings and linguistic structure. Fril incorporates a complete Prolog system which can take advantage of its built-in Definite Clause Grammar (DCG) translation and interpretation facilities to produce flexible syntactic parses and partial semantic interpretations to identify the various uncertain relations being expressed. In more difficult cases where a more semantic based approach is appropriate, a Conceptual Graph toolkit implemented in Fril and based on Sowa's theory of Conceptual Graphs [8], can be employed in conjunction with the DCG utilities.

In a previous paper [5], the authors discussed Fril modules for Fuzzy Linguistic Reasoning, based on Zadeh's PRUF: a meaning representation language for natural languages [11], but adapted to the Support Logic inference style of Fril. These adaptations included Zadeh's extensions to Test-Score Semantics and Qualitative Systems Analysis based on Fuzzy Logic [12]. In this chapter we explore the process of synthesizing sentence interpretations for fuzzy linguistic reasoning by using the DCG formalism to derive appropriate support logic queries for testing the meaning of sentences.

8.2   Fuzzy Linguistic Reasoning

Fril can express a wide range of linguistic sentences as relational knowledge with uncertainty by means of its conditional support logic rules. Three types of uncertainty rules are defined in Fril: a basic rule, an extended rule and an evidential logic rule [1]. For example, the following natural language sentence could be modelled by the associated basic rule with conditional support:

"Young wealthy business people generally drive fast cars"
((drives_car X (type fast))
       (business_person X)
       (age X young)
       (income X wealthy)) : (0.9 1)

where the support pair (0.9 1) models the term 'generally' and represents a probability interval expressing the conditional support for an arbitrary person X driving a fast car given that he/she is a young and wealthy business person. The terms fast, young and wealthy would be most appropriately modelled by fuzzy sets. For example, the following 'itype' defines a piece-wise linear fuzzy set for 'young' over the continuous domain of ages, where points between the domain:membership pairs are determined by linear interpolation:

       set (ages 0 120)
       (young [25:1, 30:0.7, 35:0.3, 40:0] ages)

Discrete fuzzy sets can also be defined. For example the following defines the fuzzy set for 'high_score' over the domain 'toss' in a fair toss of a die:

set (toss (1 2 3 4 5 6))
(high_score {4:0.2, 5:0.8, 6:1} toss)

The basic rule representation also allows pairs of support pairs to represent the conditional support for the rule head given that the body is true and the conditional support for the rule head given that the body is false. The extended rule is a straightforward generalization of this rule expressing conditional supports for the head given the truth of constituent goals in the body of a rule. This extension is particularly useful in the context of modelling causal nets [1, 7]. The evidential logic rule is applicable for expressing inductive and abductive knowledge and is particularly relevant to case-based reasoning [1, 9]. For example, the following natural language statement could be modelled by the associated evidential logic rules:

> "Comfortable chairs generally have most of the following features:
> a more or less horizontal, rectangular, soft seat surface,
> four well-spaced vertical supporting legs, about 0.3 metres long,
> a tall supporting and slightly sloping back,
> a pair of arm rests"

((comfy_chair X) (evlog most (
    (seat_surface X appropriate) 0.4
    (supporting_legs X adequate) 0.3
    (chair_back X present) 0.2
    (arm_rests X fitted) 0.1))) : ((1 1) (0 0))

((seat_surface X appropriate) (evlog most (
    (seat_slope X more_or_less_horizontal)0.5
    (seat_shape X approximately_rectangular) 0.3
    (seat_material X soft) 0.2))) : ((1 1) (0 0))

((supporting_legs X adequate) (evlog most (
    (topology_legs X well_spaced) 0.4
    (orientation_legs X about_vertical) 0.2
    (length_legs X "approx_0.3") 0.4))) : ((1 1) (0 0))

((chair_back X present) (evlog most (
    (angle_back X slightly_sloping) 0.4
    (length_back X tall_back) 0.6))) : ((1 1) (0 0))
((arm_rests X fitted)
    (feature X arm_rests pair)) : ((1 1) (0 0))

where the weights on the goals in the bodies reflect degrees of relative importance, the fuzzy set 'most' is a filter used to model a soft form of conjunction and typically might be defined as follows:

set (proportion 0 1)
(most [0.5:0, 0.8:1] proportion)

and the concepts more_or_less_horizontal, approximately_rectangular, soft, well_spaced, about_vertical, 'approx_0.3', slightly_sloping and tall_back are all defined by fuzzy sets.

In addition to the three types of rules, Fril can represent Fuzzy Control rules [1, 6] in a completely natural way by defining such rules with fuzzy sets in the heads. For example, the following extract from a set of fuzzy linguistic statements expressing control knowledge could be modelled by the given Fril basic rules:

"If the displacement is small and the speed is slow then apply minimal control"
"If the displacement is small and the speed is fast then apply intermediate control"
"If the displacement is large and the speed is fast then apply drastic control"

    set (control_space . . .)
    set (disp_space . . .) . . . *etc.*

    (minimal [. . .] control_space)
    (intermediate [. . .] control_space)
    (drastic [. . .] control_space) . . . *etc.*

((control_force X minimal) (displacement X small) (speed X slow)):(0.8 1)
((control_force X intermediate) (displacement X small) (speed X fast)):(0.9 1)
((control_force X drastic) (displacement X large) (speed X fast)):(1 1)
            . . . etc.

Given measurement information on the displacement and speed of some plant, the following Fril queries obtain an expected fuzzy set output inference E on control_space and a defuzzified output control D respectively:

    qse((control_force plant E))
    qsv((control_force plant D))

Fril focuses on the modelling of uncertain inference by reasoning with support pairs, and this is in accord with the underlying theory of mass assignments which unifies fuzzy and probabilistic reasoning by showing how both probability and possibility distributions induce mass assignments [1, 4]. Fril also includes basic fuzzy arithmetic operations in the core language, as well as the partial match of fuzzy terms by semantic unification. However, since Fril incorporates a complete dialect of Prolog it is also possible to implement other modes of uncertain inference such as Bayesian probability inference and Fuzzy Logic reasoning. A Bayesian inference module has been developed for Causal Net applications and the Fril modules for Fuzzy Linguistic Reasoning [1] include various fuzzy set processing routines such as max-min composition, hedge modifiers such as 'very', so that 'very tall' can be generated from 'tall', and test-score semantics [5, 12] for fuzzy quantified propositions, based on the support pair characterization of probability and semantic unification. Some examples of these derived from fuzzy linguistic statements are as follows:

"George earns somewhat more than Sarah, who is very poor"

set (salary 0 100000)
set (truth 0 1)
(very [0.7:0, 1:1] truth)
(poor [5000:1, 10000:0] salary)
(somewhat_more [3000:0, 6000:1, 9000:0] salary)

? ((tfm very poor VP) (composing2 VP salary somewhat_more SAL)
                    (p George earns salary SAL)(pp))
George earns salary [3000:0, 6000:1, 11000:1, 15500:0]

Fuzzy quantified statements can be often be effectively modelled using Zadeh's concept of test-score semantics which is also implemented in Fril. For example:

"Most obese people eat mostly rich foods"

((eats_mostly_rich_foods (X) REL) (testscore_relation most ((food Y)
                    (rich Y)) are ((eats X Y)) by (X) REL))

where REL is inferred as unary support logic relation on the space of people, characterizing the degree to which each person in the given database, eats mostly rich foods. The relevant query is then:

? ((eats_mostly_rich_foods (P) R)
      (test_score most ((weight P obese)) are ((R M)) by () S)
      (p Test score for most obese people eating mostly rich foods is
                    support pair S) (pp))
Test score for most obese people eating mostly rich foods is
                    support pair (0.64 0.89)

"In recent years, young dealer Nick made much more money than
                    most of his older colleagues"

This can be treated in a similar fashion to the above example by interpreting the sentence as: most 'od' are 'mm', where 'od' and 'mm' are relations defined as follows: od(x) denotes 'dealer x is older than Nick, who is young'; mm(x) denotes 'Nick made much more money than x in recent years'.

((od X)
      (? ((age Nick _AgeNick) (age X _Age) (sum _Diff _AgeNick _Age)))
      (match young _AgeNick)
      (match "somewhat_larger_than_zero_age" _Diff)) : ((1 1)(0 0))

((mm X)
      (compute_recent_earnings Nick _EarnNick)
            (compute_recent_earnings X _Earn) (sum _Diff_Earn_EarnNick)))

(match "much_larger_than_zero_earnings" _Diff)) : ((1 1)(0 0))

Then the appropriate query is as follows:
> ? ((test_score most ((od X)) are ((mm X)) by () S)
>   (p Support is S) (pp))
> Support is (0.85 0.97)

In this example, the concept of a test score involves finding the proportion of dealers satisfying relations 'od' and 'mm' counted over those satisfying 'od', and testing this proportion (expressed as a support pair) against the fuzzy quantifier 'most'.

8.3   Sentence Interpretation by Linguistic Processing

Each of the examples of linguistic reasoning which have been modelled above involves the translation from a natural language statement into a suitable pseudo-natural language structural form which is the basis of the relevant FRIL query. For many problems, the rules of translation are simple, intuitive and readily mechanized. In such cases, such mechanization can easily be implemented in FRIL, for example using the built-in Definite Clause Grammar translation and interpretation facilities of the language. As a simple example of linguistic processing, consider the following Definite Clause Grammar (DCG) fragment for deriving support for the test-score semantics goal from the following linguistic statement:

> "Most young dealers are rich"

> ((test_score_support S) →
>   (quantifier _Quant)
>   (noun_phrase W _Antecedent)
>   (verb_phrase W _Consequent)
>   (? ((test_score _Quant _Antecedent are _Consequent by () S))))

> ((quantifier _Quant) → ((Q)) (? ((fuzzy_quantifier Q _Quant))))

> ((noun_phrase W (_NounGoal _AdjGoal)) →
>   (adjective W _AdjGoal)
>   (noun W _NounGoal))
> ((noun_phrase W (_AdjGoal)) →
>   (adjective W _AdjGoal))

> ((verb_phrase W _Consequent) →
>   (verb)
>   (noun_phrase W _Consequent))

((adjective W (_Pred W _FuzzAdj)) →
    ((_Adj))
    (? ((fuzzy_adjective _Adj _FuzzAdj) (get_pred _Adj _Pred))))

((noun W (_Pred W)) →
    ((_Noun))
    (? ((get_root _Noun _Root) (get_pred _Root _Pred))))

((verb) → ((are)))

((fuzzy_quantifier most fuzmost))

((fuzzy_adjective young fuzyoung))
((fuzzy_adjective rich fuzrich))

((get_root dealers dealer))

((get_pred young age))
((get_pred rich wealth))
((get_pred dealer dealer_name))

This grammar can be used together with an appropriate knowledge base of which the following could be a fragment:

set (agedom 0 120)
set (wealthdom 0 1000)
set (probdom 0 1)
(fuzyoung [25:1, 30:0.9, 40:0] agedom)
(fuzrich [50:0, 150:0.9, 200:1] wealthdom)
(fuzmost [0.5:0, 0.8:1] probdom)

((age Nick 28))
((age Bruce 43))
((age Ted fuzyoung))
((age Brian 32))
. . .
((dealer_name Nick))
((dealer_name Ted))
((dealer_name Brian))
. . .
((wealth Nick 500))
((wealth Ted 170))
((wealth Brian 185))
. . .

The following goal can then be called to derive the corresponding query:

? ((test_score_support S (most young dealers are rich) ()) (pp S))
(0.78 0.85)

where this support is derived from the execution of the following test-score semantics goal which is obtained from the parse of the sentence (most young dealers are rich):

(test-score fuzmost ((dealer_name W) (age W fuzyoung))
                      are ((wealth W fuzrich)) by () S)

Of course, the grammar fragment displayed above is targeted specifically to the given sentence, but it can be generalized to a wide variety of related sentences of similar form. Moreover, different parses can be derived for the different contexts in which the sentence is being examined. For example, the above example tests the given sentence against a database of relevant information. Alternatively, the sentence might express heuristic knowledge which could be parsed as a Support Logic rule to be added to the knowledge base. The user interface for entering the required sentence can be made more natural and user friendly using a dialogue interface which is built into Fril.

References

[1] Baldwin J.F., Martin T.P. and Pilsworth B.W. (1995) *Fril - Fuzzy and Evidential Reasoning in Artificial Intelligence.* Research Studies Press Ltd., Taunton, UK.

[2] Baldwin J.F. (1987) Evidential Support Logic Programming. *Fuzzy Sets & Systems* 14: 1–26.

[3] Baldwin J.F., Martin T.P. and Pilsworth B.W. (1991) FRIL: A support logic programming system. *Proceedings of AI and Computer Power: The impact of Statistics*, Unicom Seminars, 159–172.

[4] Baldwin J.F. (1993) Fuzzy, Probabilistic and Evidential Reasoning in Fril. *IEEE Proc. Fuzzy Control*, San Francisco, 1–8.

[5] Baldwin J.F. and Pilsworth B.W. (1993) Fril Modules for Fuzzy Linguistic Reasoning. *Proc. EUFIT '93, First European Congress on Fuzzy and Intelligent Technologies*, Aachen, Germany, Sept. 7-10 1993, 618–623.

[6] Kosko B. (1992) *Neural Networks and Fuzzy Systems — A Dynamical Systems Approach to Machine Intelligence.* Prentice-Hall.

[7] Pearl J. (1988) *Probabilistic Reasoning in Intelligent Systems.* Morgan Kaufmann.

[8] Sowa J.F. (1984) *Conceptual Structures — Information Processing in Mind and Machine.* Addison-Wesley.

[9] Waltz D.L. (1990) Memory based reasoning. In Arbib M.A. and Robinson A. (eds), *Natural and Artificial Parallel Computing*, MIT Press.

[10] Zadeh L.A. (1965) Fuzzy Sets. *Information and Control* 8: 338–353.

[11] Zadeh L.A. (1978) PRUF — A meaning representation language for natural languages. *Int J. Man-Machine Stud.* 10: 395–460.

[12] Zadeh L.A. (1992) Knowledge Representation in Fuzzy Logic. In Yager R.R. and Zadeh L.A. (eds), *An Introduction to Fuzzy Logic Applications in Intelligent Systems*. Kluwer Academic.

9

Customer Segmentation for Banks and Insurance Groups with Fuzzy Clustering Techniques

R. Weber

9.1 Introduction

This chapter describes a project which is currently under investigation with a German financial institute. It deals with customer segmentation in the area of banking and finance.

In Section 9.2 problems and potentials of customer segmentation are discussed. Section 9.3 lays down the fundamentals of Fuzzy Cluster Analysis and introduces the idea of Neural Networks. In Section 9.4 the configuration and realisation of Fuzzy Cluster Analysis for customer segmentation is shown and results presented. Section 9.6 discusses future prospects.

9.2 Customer Segmentation — Problems and Potential

Customer segmentation has its origin in the field of marketing and market research [6]. It provides analytical division of all potential customers in a sales market according to different criteria. This results in the formation of internally homogeneous and externally heterogeneous groups of customers

or customer segments, thus providing marketing activities focused on these
segments.

9.2.1   Customer Segmentation in Financial Services

The most important component for success in financial services is certainly
the relationship between the respective institute and its customers. This
relationship and the customer's degree of satisfaction and confidence is of
central importance since it concerns the customer's property. It determines
the strategic aim in banking and finance. The needs of individual customers
are crucial for strategic planning in this area [8].

In addition to the necessity of providing an enormous variety of products,
most financial institutes serve a broad palette of customers. Due to this, financial institutes have to adapt as quickly as possible to changing requirements
and structures in order to remain in a strong position among a rising number
of competitors.

Confronted with this, financial institutes have realized that a decisive approach for the improvement of quality in consultation and services lies in the
segmentation of all customers into different target groups containing "similar"
customers. The products and services for a certain customer segment must be
provided in a way and at a level which the customer belonging to this segment
is expecting and willing to pay for. The relevance of a customer-oriented segmentation and the identification of different customer segments by suitable
methods is required for such an approach.

9.2.2   Traditional Solutions and their Limits

Traditional crisp (non-fuzzy) cluster analysis is the method usually used for
customer segmentation in market research. Cluster analysis refers to a group
of methods for the recognition of structure in data sets. It can be applied if
certain features of different objects (*e.g.* customers) can be observed, and if
those objects are supposed to be divided according to their individual feature
values. The entire set of objects will be clustered regarding all describing
features in such a way that all objects belonging to one cluster are (possibly)
similar. On the other hand, the objects belonging to different clusters should
be different regarding their feature values.

By making use of cluster analysis certain customer segments (out of the
entire set of all customers) will be discovered. Each customer will be assigned
crisply and completely to a certain cluster. Those clusters are separated from
each other by hard and precise thresholds. If the feature values of a certain
customer overstep these thresholds, this customer is assigned to a different
cluster. If the feature values of a customer reach exactly the determined
threshold, this customer is assigned optionally to one or the other cluster
completely, though he or she should belong to both clusters to the same degree.
Further, a customer whose feature values come very close to a threshold but
cannot reach it will be assigned to 'his' cluster crisply and completely, though

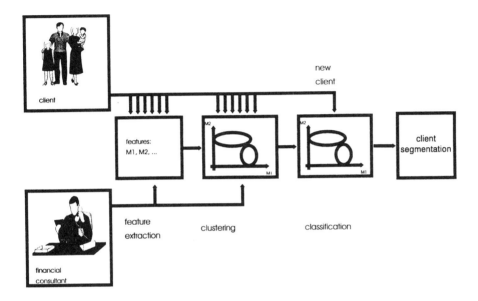

Figure 9.1: Process of customer segmentation.

he or she could nearly belong to a certain degree to another cluster beyond the threshold.

This reveals the limits of traditional cluster analysis. A given data set cannot be clustered accurately, since the method is based on crisp thresholds. In the following improved methods will be shown.

9.3   Methods for Improved Segmentation using Fuzzy Data Analysis

In the previous section the problem formulation of customer segmentation and the limitation of traditional solutions have been shown. Here we present methods and tools which offer support for financial institutes being faced with the problem of customer segmentation. These methods are based on Fuzzy Logic.

9.3.1   Basic Principles of Fuzzy Logic

The basic idea of Fuzzy Set Theory is that — in contrast to classical set theory — an element can belong to different sets with different degrees of membership, respectively [9].

Most applications of fuzzy techniques are to be found in the area of industrial control, where vague human knowledge has been used to control technical processes, *e.g.* [10]. But to an increasing degree, fuzzy methods are applied to different areas and represent powerful techniques to solve not only technical problems. Here fuzzy data analysis is one very promising field where many successful applications have been performed [2].

9.3.2   Data Analysis Based on Intelligent Techniques

The term "intelligent techniques" summarizes approaches from fuzzy logic and neural networks. These two techniques can be used to extract information from a given set of data [5]. In this chapter fuzzy cluster techniques are used to build segments of homogeneous customers. In particular, the fuzzy c-means algorithm (FCM) is applied (see *e.g.* [1]). This algorithm assigns objects, which are described by several features, to fuzzy classes. Objects belong to these classes with different degrees of membership.

The concept of neural networks aims at modelling human intelligent behaviour by imitating its information processing with computers. Different learning strategies (for example, supervised and unsupervised techniques) for neural networks have been suggested. The Kohonen algorithm presents one method for unsupervised learning [3] and leads to crisp cluster results. Recent developments have combined the theories of fuzzy logic and neural networks. In the project described in this paper the Fuzzy Kohonen Clustering Network which leads to fuzzy clusters [7] has been studied. Results of this method applied to the problem of customer segmentation will be presented in the following sections.

9.3.3   DataEngine — A Software-Tool for Intelligent Data Analysis

The software-tool DataEngine contains methods for data analysis which are based on the above intelligent techniques [4]. The combination of pre-processing, statistical analysis, and intelligent classification systems leads to a powerful software tool which can be used in a very broad range of applications.

Amongst other methods, DataEngine provides both fuzzy c-means and fuzzy Kohonen clustering networks and has been used to solve the problem of customer segmentation. This tool for intelligent data analysis is implemented as an object-oriented concept in C++ and runs on all the usual hardware platforms [4]. Interactive and automatic operation supported by an efficient and comfortable graphical user interface facilitates the application.

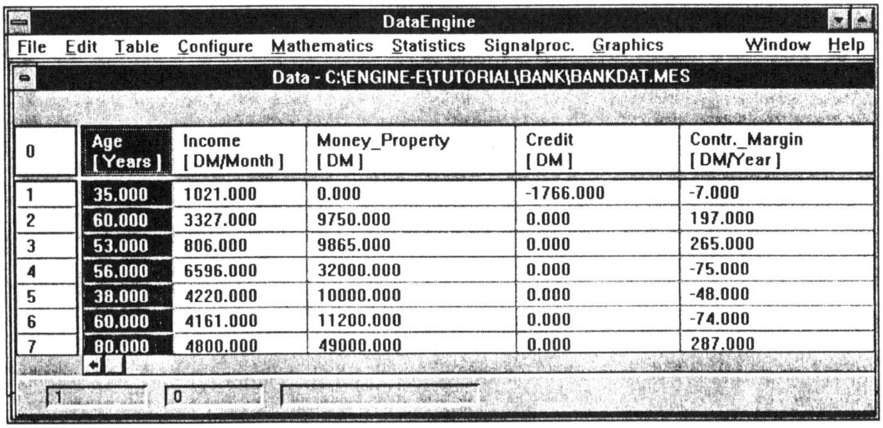

	Age [Years]	Income [DM/Month]	Money_Property [DM]	Credit [DM]	Contr._Margin [DM/Year]
0					
1	35.000	1021.000	0.000	-1766.000	-7.000
2	60.000	3327.000	9750.000	0.000	197.000
3	53.000	806.000	9865.000	0.000	265.000
4	56.000	6596.000	32000.000	0.000	-75.000
5	38.000	4220.000	10000.000	0.000	-48.000
6	60.000	4161.000	11200.000	0.000	-74.000
7	80.000	4800.000	49000.000	0.000	287.000

DataEngine — File Edit Table Configure Mathematics Statistics Signalproc. Graphics Window Help — Data - C:\ENGINE-E\TUTORIAL\BANK\BANKDAT.MES

Figure 9.2: Selected features.

9.4 Configuration and Results of the Segmentation and Consequences for Marketing Activities

In this project customer-specific data which are given by a German financial institute will be investigated for segmentation. First, the features for segmentation will be selected and examined for correlation.

The selection of the relevant features for the segmentation is of high importance for cluster analysis, because this step definitely determines which criteria lead to the final clustering of the customer data. There is an enormous variety of features available. The features can be divided into demographic features, such as age, sex and family status, and into socio-economic features such as education, profession, income and property. The finally selected features must fulfil some requirements. They should show sufficient discriminatory potential without correlation. Further, they must be quantifiable (for the application of mathematical methods) and should have similar scale-level.

After extraction of the features which are relevant for segmentation, data pre-processing will be done.

Correlation analysis should prove that the selected features are largely independent and that no attribute will be over-weighted in the cluster analysis.

Further, normalization makes the data comparable, since originally the

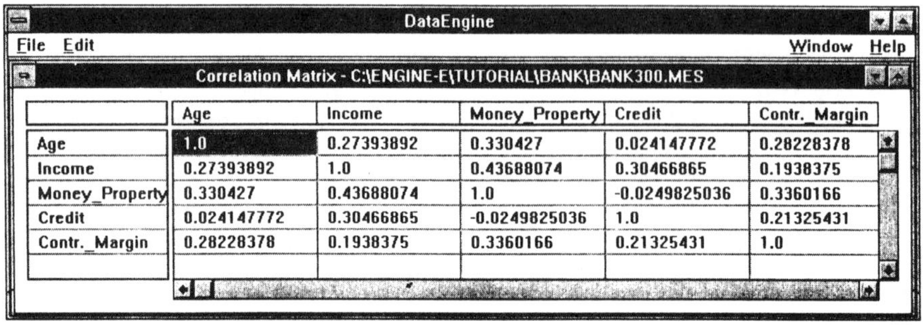

	Age	Income	Money_Property	Credit	Contr._Margin
Age	1.0	0.27393892	0.330427	0.024147772	0.28228378
Income	0.27393892	1.0	0.43688074	0.30466865	0.1938375
Money_Property	0.330427	0.43688074	1.0	-0.0249825036	0.3360166
Credit	0.024147772	0.30466865	-0.0249825036	1.0	0.21325431
Contr._Margin	0.28228378	0.1938375	0.3360166	0.21325431	1.0

Figure 9.3: Result of the correlation analysis.

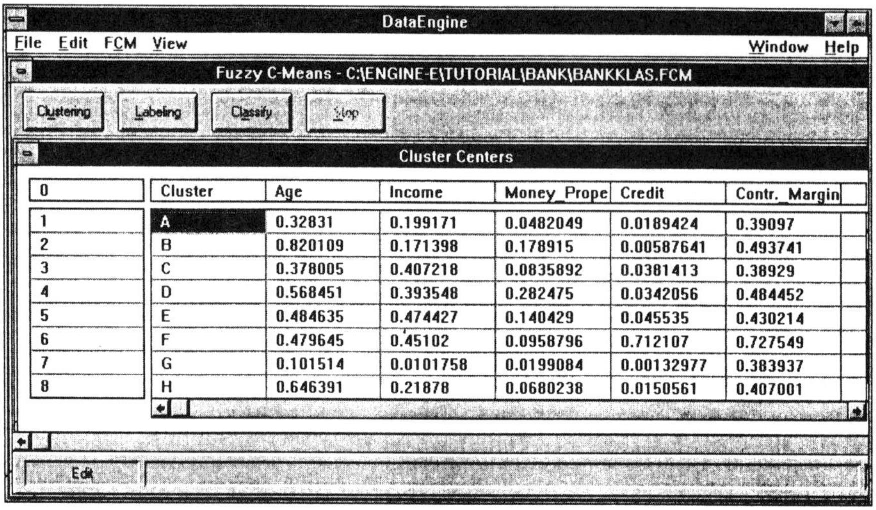

0	Cluster	Age	Income	Money_Prope	Credit	Contr._Margin
1	A	0.32831	0.199171	0.0482049	0.0189424	0.39097
2	B	0.820109	0.171398	0.178915	0.00587641	0.493741
3	C	0.378005	0.407218	0.0835892	0.0381413	0.38929
4	D	0.568451	0.393548	0.282475	0.0342056	0.484452
5	E	0.484635	0.474427	0.140429	0.045535	0.430214
6	F	0.479645	0.45102	0.0958796	0.712107	0.727549
7	G	0.101514	0.0101758	0.0199084	0.00132977	0.383937
8	H	0.646391	0.21878	0.0680238	0.0150561	0.407001

Figure 9.4: Normalized cluster centres for $c = 8$ clusters.

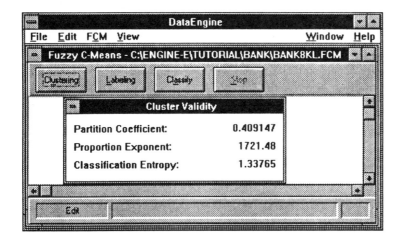

Figure 9.5: Cluster validity measures for $c = 8$ clusters.

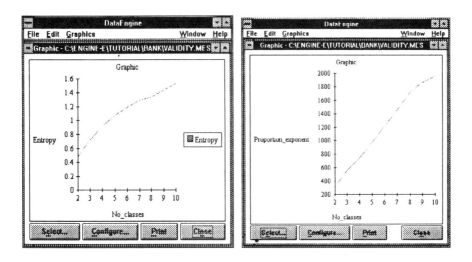

Figure 9.6: Cluster validity measures for $c = 2$ to $c = 10$ clusters.

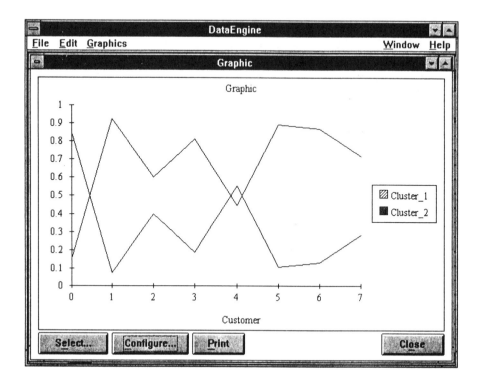

Figure 9.7: Membership values to two clusters for eight customers.

data may have different dimensions and spans of scale, which leads (without normalization) to different weights of features in the cluster-process.

9.5   Results and Benefits of Fuzzy Data Analysis for Customer Segmentation

After having determined the parameters for fuzzy c-means as well as for the fuzzy Kohonen clustering network the number of clusters is varied. For each number of clusters between $c = 2$ and $c = 10$ a fuzzy data analysis is performed. The membership values of all customers to the clusters 1 to c are computed and shown. Further, the centres of all c clusters are given.

Cluster validity measures are used to evaluate different cluster solutions. Cluster validity measures give a mathematical hint for the "best" number of clusters, which is to be found. In subsequent interviews with banking experts involved in this project, the results of the customer segmentation project described here are verified.

In contrast to the static point of view as a result of the traditional cluster techniques, fuzzy cluster analysis gives continuous degrees of membership of objects to the clusters. For each customer, membership values in all classes will be computed, which leads to a more differentiated position of each customer among the classes. Customers, for example, who show feature values of different segments will be treated with those marketing activities according to their membership values. The marketing expert of the financial institute is thus able to offer his or her customers the most fitting products and corresponding services.

In addition to this, fuzzy data analysis gives advantages regarding dynamic aspects. Based on the change of membership values to different classes, the development of a customer over time will be immediately recognized. Periodic investigations of customer data can give early hints for the movement of a customer among the classes, and as a consequence, for suitable marketing activities.

9.6   Future Prospects

Future prospects of our methods for customer segmentation arise in other fields of application which have the need for segmentation. Problems of segmentation in the area of banking and finance occur in insurance and trading companies as well. The huge amount of customer data contains a lot of information which can be extracted to provide better results in generating target groups and which allows for more customer-oriented marketing activities.

The integration of techniques for data analysis, which are described in this chapter, into standard tools for data mining and data warehousing, offers high future potential.

Recent developments in the above-mentioned project for a German financial institute have shown how the combination of such standard tools with fuzzy logic and neural networks improves the process of customer segmentation.

References

[1] Bezdek J.C. (1981) *Pattern Recognition with Fuzzy Objective Function Algorithms*. Plenum Press, New York.

[2] Bezdek J.C. and Pal S.K. (1992) (eds) *Fuzzy Models for Pattern Recognition*. IEEE Press, New York.

[3] Kohonen T. (1989) *Self-Organization and Associative Memory*. 3. Aufl., Springer-Verlag, Berlin.

[4] MIT (1995) *DataEngine 1.2. User Manual*, Aachen.

[5] Meier W., Weber R. and Zimmermann H.-J. (1994) Fuzzy Data Analysis — Methods and Industrial Applications. *Fuzzy Sets and Systems* 61: 19–28.

[6] Smith W.R. (1956) Product Differentiation and Market Segmentation as Alternative Marketing Strategies. *Journal of Marketing* 21: July, 3–8.

[7] Tsao E. C.-K, Bezdek J.C. and Pal N.R. (1994) Fuzzy Kohonen Clustering Networks. *Pattern Recognition* (27) 5: 757–764.

[8] Wood D.R. Jr. (1980) Long Range Planning in Large United States Banks. *Long Range Planning* 91–98.

[9] Zadeh L.A. (1965) Fuzzy Sets. *Information and Control* 8: 338–353.

[10] Zimmermann H.-J. (1991) *Fuzzy Set Theory - And Its Applications*. 2nd rev. ed., Kluwer, Boston.

10

Tuning of Fuzzy Controllers: Case Study in a Column Flotation Pilot Plant

M.T. Carvalho, F.O. Durão, P.J. Costa Branco

10.1 Introduction

Column flotation is a complex process aiming the concentration of minerals [6]. Due to its highly non-linear and multivariable nature, its control is a challenge to process and control engineers. Among the multiple variables involved, three of them are mandatorily controlled for the stabilization of the column operation (see Figure 10.1): LEVEL of interface between froth and collection zone, air HOLDUP (volumetric fraction of air, as air bubbles, inside the collection zone), and BIAS water flow rate (difference between volumetric flow rates of the reject and feed streams).

The solution that is usually taken in industrial plants is to pair each one of the above controlled variables with REJECT, AIR and WASHING WATER flow rates, respectively, and to implement PID controllers in each control loop. However, due to the unavoidable interaction between the control loops and to the significant measurement errors, the stabilizing control, when achieved, is rough [1] requiring the presence of a skilled operator for supervision task.

Column flotation process is therefore a candidate for fuzzy control and, in 1993, a preliminary fuzzy controller was developed [3]. The evolution of the controller was made by trial-and-error testing on a pilot plant (Figure 10.2) and the tuning was subjectively done based on some general rules found in

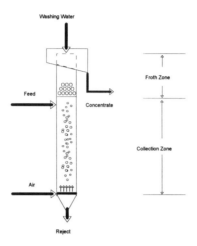

Figure 10.1: Column flotation schematic diagram.

the technical literature, *e.g.* [7], on the knowledge of the process and mainly on the team expertise. The controller was a 'classical' fuzzy controller. The fuzzification stage used sinusoidal type membership functions for all the variables. Three fuzzy sets (attributes) described each variable and the rule base, amounting to 27 rules per output variable, was heuristically defined. The rules were of the type:

IF	(Level is *Medium* and BIAS is *High* and HOLDUP is *Low*)
THEN	(REJECT is *Medium*)

Some deterministic relationships, mainly related to the process steady-state behaviour, were also used. The inference method used was the *max-prod* one and the logical connective *and* was modelled by the *min* operator. The defuzzification was performed by the gravity centre method.

Some improvements were made since then. The main changes, consequence of a better knowledge of the process, were related to the control strategy and control parameters. While in the former controller it was applied the strategy commonly used on the classical control: to pair variables and to establish all the possible combinations between input variable attributes, in the new one the flexibility that fuzzy control allows in the design of the rule base is used. For instance, when the LEVEL error is small the WASHING WATER flow rate is manipulated for LEVEL error elimination and the REJECT is manipulated for BIAS error elimination. But, when LEVEL error is large, **both** of these output variables are manipulated in order to have a faster LEVEL error correction. This strategy has advantages because, in process terms, there is no sense to control BIAS when the LEVEL is not stabilized. Therefore, this variable must be priority controlled and BIAS is controlled only when LEVEL error is small. Another improvement consisted in the indexing of the universe of discourse (UD) limits of each input variable to the respective set-point allowing for on-line variations of the input variable set-point (SP). A real-time graphics interface was provided allowing the on line visualization of all the variables and the on line changes of input variables SP and FEED flow rate [4].

As a consequence of the above-mentioned work, the idea of how easy is to design a fuzzy controller conducting to satisfactory results even for this complex process was confirmed [8].

This chapter extends the subjective tuning already done with a more systematic study. The controller sensitivity to changes of some of its parameter is investigated as well as its robustness to disturbances. Although, this kind of studies is common in several published works, *e.g.* [2], they are not oriented towards the study of real processes, neither at industrial scale nor at pilot scale.

10.2   Test Program

An extensive test program amounting to 22 tests was carried out in order to study the sensitivity of the controller to the following parameters:

- 3 discretization intervals (equal for all the variables);
- 3 UD ranges for the input and 2 for the output variables;
- 2 types of membership functions: sinusoidal and trapezoidal;
- 3 sinusoidal curve shapes: more or less fuzzy characterization of attributes;
- 3 rule base: two different control strategies and random rules elimination;
- 2 operators for intersection of input variables attributes fuzzy sets: *min* and *prod*;
- 2 operations for the union of individual rules fuzzy sets: *max* and *sum*;
- 2 defuzzification methods: gravity centre and a simplification of *Takagy-Sugeno* fuzzy inference system.

One of the main points in the design of a fuzzy controller is the number of attributes to be defined for each variable. More attributes mean more complexity in the rule base definition. As the rule base was heuristically defined, it was decided to have a simpler but more soft rule base, using only three attributes per variable.

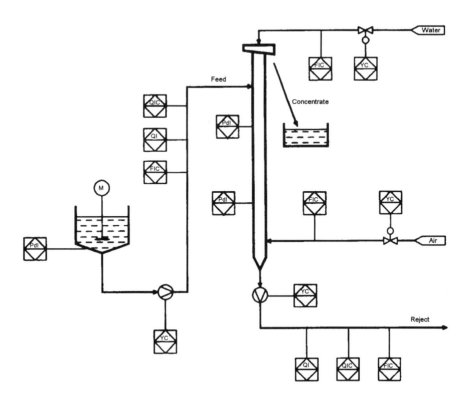

Figure 10.2: Pilot plant diagram (legend according to ISO system).

10.3 Experimental Work

The experimental work was done in the pilot plant using a two phase air/water system. All the sensors and actuating devices that have been used are represented in Figure 10.2. The HOLDUP (%) is estimated from measures of pressure cells (PdI) mounted on the column wall. The LEVEL (cm) is calculated from the measure of one pressure cell and from the HOLDUP estimation. The BIAS (l/h) is the difference between FEED and REJECT flow rates that were measured by electromagnetic flow meters, as a fraction of the FEED flow rate. All the output variables are controlled, at a lower level by PID controllers actuating on the speed of peristaltic pumps or control valve opening. A suitable dosage of frother, controlled manually, is continuously added to the system to achieve the desired HOLDUP.

The experimental strategy undertaken was to disturb the process, after the steady-state condition was reached. In all tests, step disturbances were applied to the following variables:

- FEED flow rate (this variable is one of the disturbance variables in
the industrial process; in the pilot plant it is assumed that it can
reflect all the feed characteristics disturbances that can occur in an
industrial plant);

- LEVEL and BIAS SP (in cases where it was considered necessary
a positive and a negative step on LEVEL SP were tested).

The disturbance steps performed were feasible maximum values in normal
operation: 20 l/h on FEED, 10cm on LEVEL and 5% on BIAS. The evaluation of the controller performance to the HOLDUP SP disturbance does not
make sense because this variable is very much dependent on the frother concentration inside the collection zone. Therefore, to make a step disturbance
on the HOLDUP SP it is also necessary to change the frother concentration.
The sampling time, for all variables, was about one second.

10.4   Results and Discussion

The analysis of the results, to parameter changes, was done comparing the
controller performance with that obtained for the following parameter values:

- discretization: 100 intervals;

- fuzzification: sinusoidal curves for all variables;

- rule-base with 15 rules;

- fuzzy inference: *max-prod* with *min* operator for logical connective
and;

- defuzzification method: gravity-centre.

Only the significant results are presented in Figures 10.3 to 10.15. These
represent the transient response of the input variable under study to a step
disturbance. The results are displayed each five sampling interval. For each
test, the results for different conditions of the tested parameter are plotted
in the same figure. For a better comparison some of the results were shifted
vertically.

Generally, perturbations of FEED did not affect significantly the controller robustness (except on cases presented in Figure 10.3a and 10.11). BIAS
doesn't react significantly to the disturbances introduced (except in cases presented in Figures 10.9, 10.10, 10.11 and 10.15). The variable more sensitive to
the disturbances was the LEVEL, a variable that must be priority controlled.
This is the reason why the figures are, almost all, plots of this variable.

The results of all tests were analyzed taking into account the sensor's
accuracy (for instance, 1 cm variation in LEVEL has no meaning) and that
an oscillatory behaviour will lead to shorter actuators lifetime.

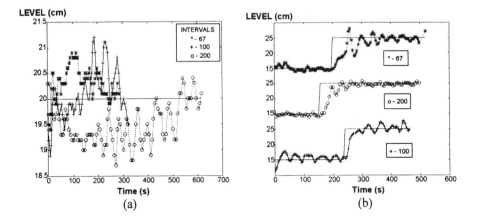

Figure 10.3: LEVEL response as a function of discretization intervals. (a) FEED perturbation (b) LEVEL SP perturbation.

10.4.1   Fuzzification

Discretization

The digital processing requires the discretization of the UD. The number of intervals to be implemented must be a trade-off between the processing time and the accuracy obtained. Processing time is, nowadays, a minor problem for slow processes like column flotation due to the increased speed of digital processors available. Three tests were performed with 67, 100 (reference test) and 200 intervals. With 200 intervals the response is slower but more attenuated and without overshoots (see Figure 10.3a). With 67 intervals the response presents an overshoot, an oscillatory behaviour and a smaller settling time compared with the results obtained for 100 intervals. For small discretizations, the membership values change more abruptly giving rise to this behaviour (see Figure 10.3b).

UD Ranges

Input variables. Initially, the UD of all three variables were symmetric with respect to the LEVEL SP. However, the LEVEL SP tracking has a behaviour that is not symmetric. The LEVEL decreases more easily than it increases. This is because when more WASHING WATER is added to increase the LEVEL (as well as the REJECT being decreased) more of that water is by-passed to the concentrate stream with no effect on the LEVEL. One way of overcoming this behaviour is to use a non-symmetric UD for this

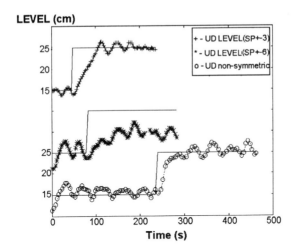

Figure 10.4: LEVEL response as a function of LEVEL UD range. LEVEL SP perturbation.

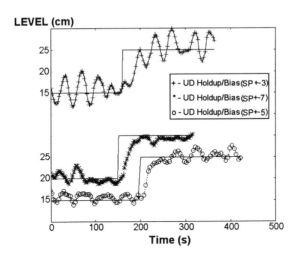

Figure 10.5: LEVEL response as a function of the BIAS and HOLDUP UD range. LEVEL SP perturbation.

variable. The behaviour to a non-symmetric UD (SP+3 as upper limit and SP-6 as lower limit) was, therefore, compared with that for a symmetric UD (SP±3 and SP±6). However looking at Figure 10.4, it can be concluded that the symmetric, but shorter, UD of SP ±3 performs slightly better than the non-symmetric UD, while with an UD of SP ±6 the control is not achieved.

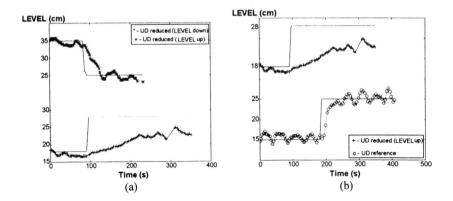

Figure 10.6: LEVEL response as a function of manipulated variables UD range. (a) Negative step perturbation on LEVEL SP, (b) Positive step perturbation on LEVEL SP.

As the HOLDUP and BIAS variables show symmetric behaviour, only symmetric UD ranges were tested: SP ±3, SP ±7 and SP ±5 (reference). Figure 10.5 shows a more oscillatory behaviour for a smaller UD range because, in this case, the output variables react in a more abrupt way, acting like a *relay*. Furthermore, when the HOLDUP UD is short, AIR is continuously actuated. As LEVEL is very sensitive to AIR, its oscillations become larger.

Output variables. The UD limits of output variables are commonly equal to the maximum limits of the actuating devices (reference test). The narrowing of the UD range was tested. As expected, this is not favourable. Figure 10.6a shows that the transient response is different for negative (down) and positive (up) step changes of LEVEL SP. In the first case, the LEVEL response to the disturbance was fast. In the other case (Figure 10.6b), as the WASHING WATER UD range was reduced, the REJECT UD range must be enlarged to satisfy the mass balance. However, a consequence of this broader REJECT UD range is the increase in BIAS oscillations amplitude that is not desired.

Types of membership functions

Fuzziness of curves. The effect of the membership function shape on the controller performance, in complex processes, is difficult to determine *a priori*. To test this parameter, sinusoidal curves were used to model the membership functions. Three shapes of curves were tested: curves more or less fuzzy compared with the reference ones (Figure 10.7). A first test was

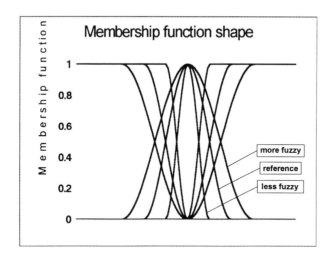

Figure 10.7: Sinusoidal membership functions shape.

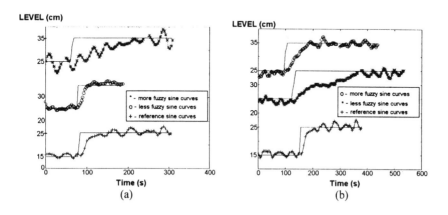

Figure 10.8: LEVEL response as a function of curve shapes. (a) LEVEL
SP perturbation (input variables), (b) LEVEL SP perturbation (output variables).

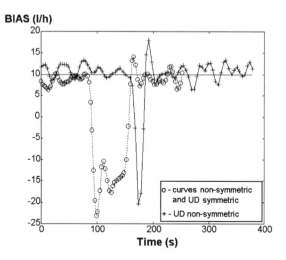

Figure 10.9: BIAS response to LEVEL SP perturbation.

carried out for the input variables. The output variables maintained the reference shape of the curves (see Figure 10.8a). A second test was done for the output variables while the input variables kept the reference curves (see Figure 10.8b). The less fuzzy curves lead to slower transient responses in both cases, and a more oscillatory behaviour when they were used to characterize input variable attributes. The more fuzzy curves give a slower response, when used on output variables, but results as good as those obtained when the input variables used the reference curves.

To adapt the controller to the above-mentioned non-symmetric behaviour of LEVEL, a non-symmetric UD or a non-symmetric membership functions for attributes characterization can be used. The comparison between these two alternatives was done and the responses for all the variables were almost the same except for BIAS when LEVEL was disturbed. The BIAS is less disturbed when a non-symmetric UD is applied (Figure 10.9).

Sinusoidal versus trapezoidal curves. Trapezoidal/triangular curves are more commonly used with fuzzy hardware than sinusoidal or other type of smooth curves. The controller performance was evaluated using trapezoidal/triangular curves and sinusoidal curves, being the shape of both types of curves the same. Sinusoidal curves, as can be seen on Figure 10.10, perform better than the former ones due to its inherently non-linearity, which is advantageous to the process.

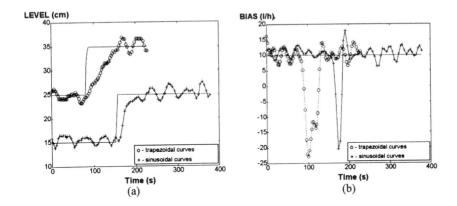

Figure 10.10: LEVEL and BIAS response as a function of the curves type. (a) LEVEL SP perturbation, (b) LEVEL SP perturbation.

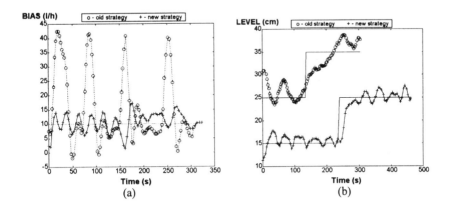

Figure 10.11: BIAS and LEVEL response as a function of rule base modification. (a) FEED perturbation, (b) LEVEL SP perturbation.

10.4.2 Rule Base

The change of the rule base was studied comparing the results of the now enhanced controller (rule base with 15 rules per output) with the one used in

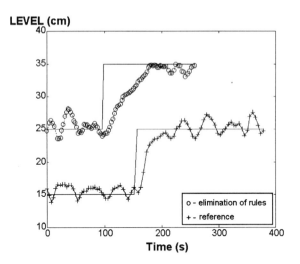

Figure 10.12: LEVEL response as a function of rule base modification. LEVEL SP perturbation.

the preliminary controller developed (with different control strategy and 27 rules). The results (see Figure 10.11) are, as expected, better with the last controller developed. The parameters of a fuzzy controller are interrelated and they are dependent on the control strategy reflected on the rule base. Therefore, when the rule base is changed (this is true for other changes too) another controller tuning would probably be necessary.

The controller robustness to elimination of rules was tested excluding randomly two and four rules per output variable (which means a decrease of 13% and 27% of the rule base). In this last case the process control was not achieved. The reason for this is that when there is a small number of linguistic attributes characterizing each variable, the fuzzy controller becomes more sensitive to the rules elimination. The elimination of two rules did not change the controller performance significantly, except for LEVEL SP perturbation (see Figure 10.12), which conducted to a slower response of LEVEL.

10.4.3   Fuzzy Inference

Fuzzy sets intersection operators

The intersection modelling was studied comparing *min* and *prod* operators (Figure 10.13). The transient response of both operators is equivalent while in the steady-state the *prod* operator provides a smoother response (as reported in several published works).

Fuzzy sets union operator

The union of individual rule fuzzy sets modelling using the *max* and *sum* operators was evaluated. The results depicted in Figure 10.14 reveal approxi-

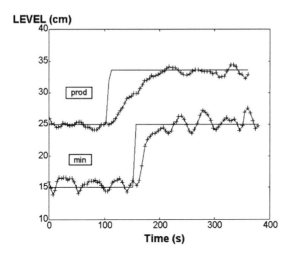

Figure 10.13: LEVEL response as a function of intersection operator. LEVEL SP perturbation.

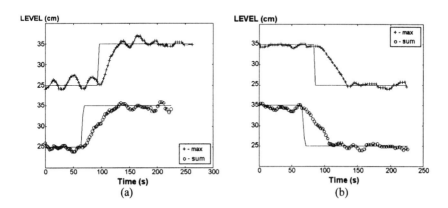

Figure 10.14: LEVEL response as a function of union operator. (a) Positive step perturbation on LEVEL SP, (b) Negative step perturbation on LEVEL SP.

mately the same behaviour with the *max* operator promoting a slightly faster transient response than the *prod* operator.

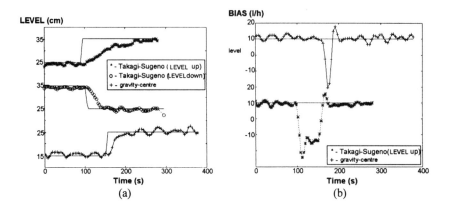

Figure 10.15: LEVEL and BIAS response as a function of the defuzzification method. (a) LEVEL SP perturbation, (b) LEVEL SP perturbation.

10.4.4   Defuzzification

The gravity-centre method was compared with a simplification of the *Takagi-Sugeno* fuzzy system. This last method is not only a defuzzification method. In its simplified form, the rules consequent is a numerical constant (interpreted as the *centroid* of the output fuzzy set) and the controller output is a weighted sum of all individual rules contribution. In general, this method is less time consuming than the former one and it is implemented in adaptive fuzzy controllers. This method leads to a slower transient response but smoother steady-state responses (Figure 10.15). The results shown come from a first trial on the numerical coefficients. However, due to the process complexity, an on-line tuning of these coefficients by training techniques would be needed.

10.5   Conclusions

The lack of well established tools and methods for tuning fuzzy controllers, that can be used in complex processes, was the main motivation for this research work. Anyone who has designed a fuzzy controller knows the difficulty of choosing, from several possibilities, the correct controller parameters. Certain rules of thumb are known and can be applied, however, sometimes if one wants to doubt some of them, some surprises can appear...

The study undertaken on the tuning of a fuzzy controller of the complex column flotation process proved to be an enormous task. A lot of work was done and more is still to be done. The main conclusion is that it is difficult to develop general rules for fuzzy controller tuning because this type of controller is very much dependent on the characteristics of the process. A complete systematic study is also impossible to do with processes that must be judged experimentally, due to the great number of tests needed.

After this study, it was concluded that the controller developed, before the tuning study reported here, based on more or less subjective reasoning rules, was good enough to control the process. The sensitivity study carried out resulted in minor corrections to the controller parameters: a decrease in LEVEL UD range, symmetric LEVEL UD and *prod* operator for intersection of fuzzy sets. The knowledge of the process and fuzzy control are the keys for a good controller development.

The *Takagi-Sugeno* method for rules definition/defuzzification is a promising technique allowing for an on-line learning strategy which is potentially advantageous when ore is added to the system, permitting the on-line controller to adapt to different ore types with different dynamic behaviour. This technique is a new research area of this team [5].

References

[1] Bergh L.G. and Yianatos J.B. (1993) Experimental studies on Flotation Column Dynamics. *Minerals Engineering* (7) 345–355.

[2] Bühler H. (1994) *Réglage par Logique Floue*, Presses Polytechniques et Universitaires Romandes.

[3] Carvalho T., Sousa J., Martins P. and Durão F. (1993) Application of Fuzzy Control to a Column Flotation Process. *Proc. of XXIV Int. Symposium APCOM* (3) 193–200.

[4] Carvalho T., Sousa J., Durão F. and Martins P. (1994) Real Time Fuzzy Control of Column Flotation Process. *Proc. of AIRTC'94* (1) 87–92.

[5] Costa Branco P.J. and Dente J.A. (1994) Inverse Model Identification Using Fuzzy-Logic – Application to an Electro-Hydraulic System, *Proc. of Int. Conf. on Electrical Machines (ICEM'94)* (2) 506–511.

[6] Finch J.A. and Dobby G.S. (1990) *Column Flotation*. Pergamon Press.

[7] Holmblad L.P. and Ostergaard J.J. (1982) Control of a Cement Kiln by Fuzzy Logic. *Fuzzy Information and Decision Processes*, Gupta M.M. and Sanchez E. (eds), 389–399, North-Holland Publishing Company.

[8] Mamdani E.H. (1993) Twenty years of Fuzzy Control: Experiences Gained and Lessons Learnt. *Proc. of Second IEEE Int. Conf. on Fuzzy Systems* (1) 339–344.

11

Exploratory Data Processing using a Fuzzy Generalization of the GUHA Approach

M. Holeňa

11.1 Introduction

The term *exploratory data processing* commonly denotes the kind of data processing aimed at the discovery of decision support and interesting relationships captured in the processed data. That kind of data processing is essential in order to improve cooperation between data and knowledge bases of decision support systems. A particular approach to exploratory data analysis is the one known under the name *General Unary Hypotheses Automaton* (GUHA). Its principle consists in automatic derivation of all interesting hypotheses supported by the data.

The method of automatic generating of hypotheses, underlying GUHA, attracted much attention in the late seventies [2, 3, 5]. First simple implementations of the method emerged very soon after [6, 8], followed later by more sophisticated systems [2, 4, 9].

Basically, the hypotheses generated by the GUHA method are sentences of an *observational calculus*, as the model of which a dichotomous *data matrix* is used, viewed as a realization of a random sample of binary vectors. Each component of the vectors, *i.e.* each column of the matrix, corresponds to some *atomic property* of real objects. Actually, only two particular kinds of sentences are considered in the method [1, 3]:

- A *GUHA implication* captures the fact that some combination of present or absent atomic properties implies another such combination. It is defined by means of an implicational quantifier and concerns the conditional probability of the random variable corresponding to the latter combination, conditioned on the random variable corresponding to the former combination.

- A *GUHA association* captures a general dependence between two combinations of present or absent atomic properties. It is defined by means of an associational quantifier and concerns the independence of the random variables corresponding to the combinations considered.

An atomic property may be, in particular, the property that some attribute has a particular value. In that way, GUHA can handle nominal data as well. In addition, it incorporates the treatment of relativized sentences and of missing data.

Most of the procedures for generating GUHA implications and associations, employed in the existing implementations, are based on statistical hypotheses testing. However, there is an *inner contradiction* inherent to using common statistical tests to this end. Those tests always require a *precise formulation* of the tested conditions on the random variables characterizing the underlying real phenomena. On the other hand, due to its explorative purpose, GUHA is predominantly used in situations in which the user has only a rather *vague knowledge* of the underlying phenomena, thus being unable to make a competent choice of the conditions to be tested. Therefore, it would be advantageous to reflect the vagueness of the user's *a priori* knowledge in the way in which sentences in GUHA are generated. To generalize the GUHA approach in that direction is the objective of the research reported in this chapter. The proposed generalization is based on a method of fuzzy hypotheses testing, which was elaborated for the statistical tests used in GUHA, and integrated into the context of an observational calculus. Due to space limitations and to the different nature of GUHA implications and GUHA associations, only the former are treated here in detail.

In the next section, the basic definitions concerning GUHA implications are recalled. In Section 11.3, fuzzy hypotheses corresponding to two GUHA implications commonly encountered in the existing implementations are proposed, and their respective test statistics are derived. The core of the chapter is formed by Section 11.4, in which a fuzzy generalization of GUHA implications is defined. The quantifiers of the two particular GUHA implications considered before are shown to be fuzzy-implicational, and several important properties of them are established. The chapter is concluded by an *overview of intended extensions*, necessary to make the approach practically applicable.

11.2   GUHA fundamentals recalled

Definition 11.1 *Let $n \in \mathcal{N}$ and $t = (t_1, \ldots, t_n) \in \mathcal{N}^n$. For each $m \in \mathcal{N}$,* *denote $\mathcal{M}(m)$, more precisely $\mathcal{M}_{\{0,1\}}(m)$, the set of all $\{0,1\}$-structures of type 1^m with finite natural domain, i.e.*

$$\mathcal{M}(m) \stackrel{\text{def}}{=} \{(S, f_1, \dots, f_{m_q}) : S \subset \mathcal{N} \ \& \ |S| < \infty \ \& \ (\forall i \in \{1, \dots, m\})$$
$$f_i : S \to \{0,1\}\}, \tag{11.1}$$

where the symbol $|S|$ stands for the cardinality of a set S.

An observational predicate calculus (OPC) of type t is given by

$$(\{P_1, \dots, P_n, =\}, X, x, J, Q), \ where$$

$\{P_1, \dots, P_n, =\}$, *is a set of* predicates, $=$ *being the identity predicate,*
X *is an at most countable set of* variables,
$x \in X$ *is a designated variable,*
$J \stackrel{\text{def}}{=} \{0, 1, \neg, \&, \vee, \Rightarrow, \Leftrightarrow\}$ *is the set of* junctors,
Q *is an at most countable set of* generalized quantifiers,
and it holds:

- *$(\forall i \in \{1, \dots, n\})$ the arity of P_i of is t_i,*

- *$(\forall q \in Q)(\exists m_q \in \mathcal{N})$ the arity of q is m_q,*

- *$(\forall q \in Q)$ a unique function $\mathrm{Af}_q : \mathcal{M}(m_q) \to \{0,1\}$ is attached to q, called the associated function of q, such that*

    a) if $M_1, M_2 \in \mathcal{M}(m_q)$ & M_1 and M_2 are isomorphic structures, then $\mathrm{Af}_q(M_1) = \mathrm{Af}_q(M_2)$

    b) the function Af defined

$$\mathrm{Af}(q, M) \stackrel{\text{def}}{=} \mathrm{Af}_q(M) \ \mid \ q \in Q, M \in \mathcal{M}(m_q)$$
$$\tag{11.2}$$

is recursive in both variables.

If in particular $t_1 = \dots = t_n = 1$, then the OPC is called monadic.

Remark 11.1 The key idea of the GUHA method is to view, for each generalized quantifier q, the elements of the set $\mathcal{M}(m_q)$ as realizations of m_q-dimensional random samples, all such random samples being defined on the same probability space. Hence the associated function Af_q of each $q \in Q$ can be viewed as a realization of a binary random variable. Since random variables expressible as compositions of random samples with functions of many variables are also commonly used as test statistics for testing statistical hypotheses, it is possible to establish a straightforward connection between generalized quantifiers occuring in observational predicate calculi, and statistical inference based on hypotheses testing. The principles of that connection will be outlined for the case of binary generalized quantifiers.

Let $(\Omega, \mathcal{A}, P)$ be a probability space, and for each $k \in \mathcal{N}$, let T_k be a test statistic on $(\Omega, \mathcal{A}, P)$ which can be expressed as a composition of a random sample of k two-dimensional binary vectors with a function of $2k$ binary variables. Finally, let $K(\alpha) \in \mathcal{B}(\Re)$ (an element of the Borel σ-algebra on reals) be a critical region for a significance level $\alpha \in (0,1)$, corresponding simultaneously to all test statistics T_k for $k \in \mathcal{N}$. Then we can define a generalized quantifier q by means of either of the following definitions of the associated function Af_q:

$$\mathrm{Af}_q(S, f_1, f_2) \stackrel{\text{def}}{=} \left\{ \begin{array}{ll} 1 & | \quad (S, f_1, f_2) \in \mathcal{M}(2) \ \& \ T_{|S|} \in K(\alpha) \\ 0 & | \quad (S, f_1, f_2) \in \mathcal{M}(2) \ \& \ T_{|S|} \notin K(\alpha), \end{array} \right. \tag{11.3}$$

or

$$\mathrm{Af}_q(S, f_1, f_2) \stackrel{\text{def}}{=} \left\{ \begin{array}{ll} 1 & | \quad (S, f_1, f_2) \in \mathcal{M}(2) \ \& \ T_{|S|} \notin K(\alpha) \\ 0 & | \quad (S, f_1, f_2) \in \mathcal{M}(2) \ \& \ T_{|S|} \in K(\alpha). \end{array} \right. \tag{11.4}$$

Consequently, testing a particular statistical hypothesis can be viewed as applying a particular generalized quantifier to formulae of an appropriate observational calculus.

Definition 11.2 *Let φ, ψ be conjunctive formulae of a monadic OPC, each of which contains only one free variable, namely the designated variable x. With such formulae, the simplified notation $\varphi \sim \psi$ instead of $(\sim x)(\varphi, \psi)$ will be used whenever $\sim$ is a binary generalized quantifier. For each $M = (S, f_1, f_2) \in \mathcal{M}(2)$, denote*

$$a_M \stackrel{\text{def}}{=} |\{s : s \in S \ \& \ f_1(s) = f_2(s) = 1\}|,$$
$$b_M \stackrel{\text{def}}{=} |\{s : s \in S \ \& \ f_1(s) = 1 \ \& \ f_2(s) = 0\}|. \tag{11.5}$$

Finally, let $\to$ be a particular binary generalized quantifier of that OPC, such that its associated function $\mathrm{Af}_\to$ fulfils

$$(\forall M_1, M_2 \in \mathcal{M}(2)) \ a_{M_2} \geq a_{M_1} \ \& \ b_{M_2} \leq b_{M_1} \ \& \ \mathrm{Af}_\to(M_1)$$
$$= 1 \Rightarrow \mathrm{Af}_\to(M_2) = 1. \tag{11.6}$$

Then the quantifier $\to$ is called implicational *and $\varphi \to \psi$ is called a* GUHA implication.

Example 11.1 Let $\alpha \in (0, \frac{1}{2})$, $\theta \in (0,1)$ be given constants. For each $M = (S, f_1, f_2) \in \mathcal{M}(2)$, denote $r_M \stackrel{\text{def}}{=} a_M + b_M = |\{s : s \in S \ \& \ f_1(s) = 1\}|$. Finally, let $\to'$, more precisely $\to'_\theta$, be a generalized quantifier defined by means of the associated function

$$\mathrm{Af}_{\to'_\theta}(M) \stackrel{\text{def}}{=} \left\{ \begin{array}{ll} 1 & | \quad M \in \mathcal{M}(2) \ \& \ \sum\limits_{i=a_M}^{r_M} \binom{r_M}{i} \theta^i (1-\theta)^{r_M - i} \leq \alpha \\ & \\ 0 & | \quad M \in \mathcal{M}(2) \ \& \ \sum\limits_{i=a_M}^{r_M} \binom{r_M}{i} \theta^i (1-\theta)^{r_M - i} > \alpha. \end{array} \right. \tag{11.7}$$

In [3] it was proven that this quantifier is implicational. In accordance with the terminology used there, a GUHA implication $\varphi \to_\theta^! \psi$ will be called a *likely implication* with a threshold θ.

Observe that (11.7) can be reformulated as (11.3) provided the corresponding test statistic is for each $M = (S, f_1, f_2) \in \mathcal{M}(2)$ defined

$$T_{|S|} = \sum_{i=a}^{r_M} \binom{r_M}{i} \theta^i (1-\theta)^{r_M-i}, \tag{11.8}$$

where a is a random variable the realization of which is a_M, and the corresponding critical region is

$$K(\alpha) = (0, \alpha). \tag{11.9}$$

Example 11.2 Let α and θ be as above. Another example of an implicational quantifier is the generalized quantifier $\to^?$, more precisely $\to_\theta^?$, which is defined by means of the associated function

$$\mathrm{Af}_{\to_\theta^?}(M) \overset{\text{def}}{=} \begin{cases} 1 & | & M \in \mathcal{M}(2) \ \& \ \sum_{i=0}^{a_M} \binom{r_M}{i} \theta^i (1-\theta)^{r_M-i} > \alpha \\ 0 & | & M \in \mathcal{M}(2) \ \& \ \sum_{i=0}^{a_M} \binom{r_M}{i} \theta^i (1-\theta)^{r_M-i} \le \alpha. \end{cases} \tag{11.10}$$

A GUHA implication $\varphi \to_\theta^? \psi$ will be called a *suspicious implication* with a threshold θ, similarly to the term used in [3], where also a proof that $\to^?$ is implicational can be found.

Different to the case of $\to^!$, the associated function of $\to^?$ can be reformulated by means of (11.4), with the test statistic defined for each $M = (S, f_1, f_2) \in \mathcal{M}(2)$ by

$$T_{|S|} = \sum_{i=0}^{a_M} \binom{r_M}{i} \theta^i (1-\theta)^{r_M-i} \tag{11.11}$$

and the critical region (11.9).

11.3 Fuzzy Hypotheses for Likely and Suspicious Implications

Recall [3] that viewing the system $(f_1(s), f_2(s))_{s \in S}$ for $M = (S, f_1, f_2) \in \mathcal{M}(2)$ as a realization of a two-dimensional binary random sample implies:

(i) Under the assumption $0 < \Pr(f_2 = 1 | f_1 = 1) < 1$, using the notation r and a for the random variables the realizations of which are r_M and a_M, respectively, the conditional distribution of a conditioned on the particular value r_M of r is binomial with parameters r_M and $p \in (0, 1)$, *i.e.*

$$(\forall i \in \{0, \ldots, r_M\}) \ \Pr(a = i | r = r_M) = \binom{r_M}{i} p^i (1-p)^{r_M-i}, \tag{11.12}$$

where

$$p = \Pr(f_2 = 1 | f_1 = 1). \tag{11.13}$$

Hence, provided $f_1 = 1$ and $f_2 = 1$ are interpreted as the observations of validity of some conjunctive formulae φ and ψ, respectively, the parameter value p can be interpreted as the conditional probability $p_{\psi|\varphi}$ of validity of ψ, conditioned on validity of φ, i.e. $p = p_{\psi|\varphi}$.

(ii) The test statistics and the critical region corresponding to the likely and suspicious implications can be used for testing statistical hypotheses concerning $p = p_{\psi|\varphi}$, in particular (11.8)–(11.9) for testing the null hypothesis

$$p_{\psi|\varphi} \leq \theta, \tag{11.14}$$

whereas (11.9) and (11.11) for testing the null hypothesis

$$p_{\psi|\varphi} \geq \theta. \tag{11.15}$$

The contradiction between the precise formulation of the tested hypotheses and the explorative character of the GUHA approach, mentioned in the introduction, also concerns the likely and suspicious implications. Consequently, it would be much more appropriate for the purpose of exploratory data analysis to replace the strictly formulated hypotheses (11.14) and (11.15) with more vague and relaxed ones. To this end, the method of fuzzy hypotheses testing proposed in [10] will be used. According to that method, a null hypothesis to be tested is viewed as a normalized fuzzy set $\tilde{\Pi}$ on a set Π of admissible parameters,

$$\tilde{\Pi} \stackrel{\text{def}}{=} \{(p, \mu(p)) : p \in \Pi\} \text{ with } \mu : \Pi \to \langle 0, 1 \rangle, \tag{11.16}$$

where $\Pi = (0, 1)$ if we are interested in null hypotheses concerning the parameter p of the binomial distribution (11.12).

The form of the membership function μ of $\tilde{\Pi}$ reflects the nature of the vagueness captured by the null hypothesis. However, a question arises how large should the involved vagueness be, i.e. how far to go when relaxing the condition (11.14) or (11.15).

Consider at first the likely implication. A particular application may suggest a particular value θ in (11.14). Then only the requirement that (11.14) should hold exactly can be relaxed. Instead, we can consider a null hypothesis paraphrased

$$p_{\psi|\varphi} \text{ is not much higher than } \theta. \tag{11.17}$$

In that case, the membership function μ of $\tilde{\Pi}$ could have, for example, one of the forms shown in Figure 11.1.

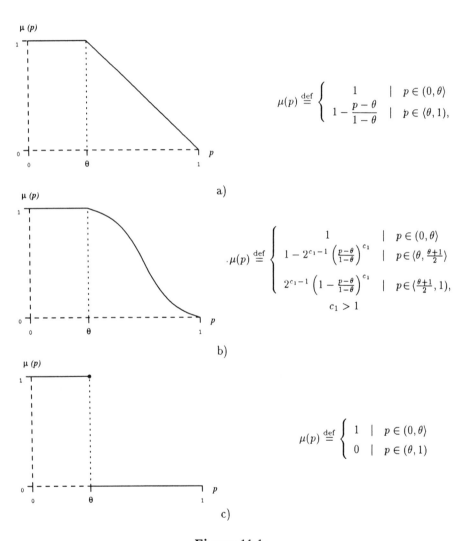

Figure 11.1:

However, very often the value θ in (11.14) is actually immaterial, and its purpose is solely to prevent $p_{\psi|\varphi}$ from being too high. In that case, we can also relax the requirement that $p_{\psi|\varphi}$ is to be compared to a particular value, and consider a null hypothesis paraphrased

$$p_{\psi|\varphi} \text{ is low.} \tag{11.18}$$

That case is illustrated by the membership functions in Figure 11.2.

The considerations that follow will not be limited to any of the above particular examples unless explicitly stated. Instead, they will only require the membership function μ corresponding to (11.18) to have the following general properties:

$$
\begin{array}{ll}
\text{(i)} & \mu \text{ is non-increasing on } (0,1), \\
\text{(ii)} & \lim_{p \to 0+} \mu(p) = 1, \\
\text{(iii)} & \lim_{p \to 1-} \mu(p) = 0.
\end{array}
\tag{11.19}
$$

In the terminology of fuzzy quantifiers [11], $\tilde{\Pi}$ with μ fulfilling the conditions (11.19) is a non-increasing linguistic quantifier. Observe that these conditions cover not only the examples in Figure 11.2, but also the examples of membership functions corresponding to (11.17) in Figure 11.1. For this reason, the fuzzy null hypothesis '$p_{\psi|\varphi}$ is not much higher than θ' will in the following be treated together with the fuzzy null hypothesis '$p_{\psi|\varphi}$ is low'.

The case of the suspicious implication can be treated in a similar way. Instead of (11.15), we can consider the null hypothesis

$$p_{\psi|\varphi} \text{ is not much lower than } \theta, \tag{11.20}$$

implying that we insist on the importance of the value θ, or

$$p_{\psi|\varphi} \text{ is high,} \tag{11.21}$$

which implies that the value θ is immaterial. Examples of membership functions corresponding to (11.20) are given in Figure 11.3, those corresponding to (11.21) are given in Figure 11.4. Their definitions are straightforward analogies of the definitions of the corresponding examples in Figures 11.1 and 11.2, they are therefore omitted.

In general, a membership function corresponding to (11.21) will only be required to fulfil an analogy of (11.19), namely

$$
\begin{array}{ll}
\text{(i)} & \mu \text{ is non-decreasing on } (0,1), \\
\text{(ii)} & \lim_{p \to 0+} \mu(p) = 0, \\
\text{(iii)} & \lim_{p \to 1-} \mu(p) = 1.
\end{array}
\tag{11.22}
$$

In the terminology of [11], $\tilde{\Pi}$ with μ fulfilling (11.22) is a non-decreasing linguistic quantifier. Since (11.22) covers also the examples of membership functions corresponding to (11.20), the fuzzy null hypothesis '$p_{\psi|\varphi}$ is not much lower than θ' will be in the sequel treated together with the fuzzy null hypothesis '$p_{\psi|\varphi}$ is high'.

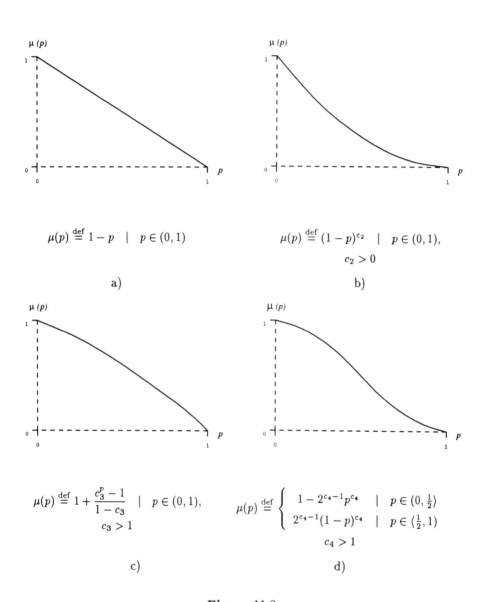

$$\mu(p) \stackrel{\text{def}}{=} 1 - p \quad | \quad p \in (0,1)$$

a)

$$\mu(p) \stackrel{\text{def}}{=} (1-p)^{c_2} \quad | \quad p \in (0,1),$$
$$c_2 > 0$$

b)

$$\mu(p) \stackrel{\text{def}}{=} 1 + \frac{c_3^p - 1}{1 - c_3} \quad | \quad p \in (0,1),$$
$$c_3 > 1$$

c)

$$\mu(p) \stackrel{\text{def}}{=} \begin{cases} 1 - 2^{c_4 - 1} p^{c_4} & | \quad p \in (0, \frac{1}{2}) \\ 2^{c_4 - 1}(1 - p)^{c_4} & | \quad p \in \langle \frac{1}{2}, 1) \end{cases}$$
$$c_4 > 1$$

d)

Figure 11.2:

Following [10], a test statistic for a fuzzy null hypothesis will be a random variable assuming values in the space of real functions on the set of parameters and expressible as a composition of a random sample of length $k \in \mathcal{N}$, with a

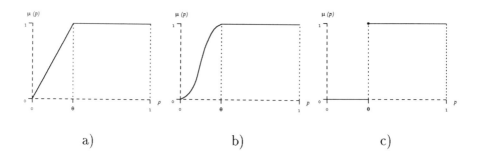

Figure 11.3:

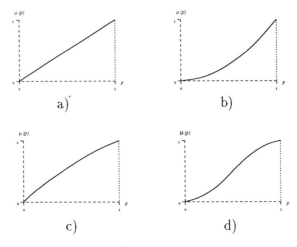

Figure 11.4:

mapping of k variables. Similarly, a critical region for a fuzzy null hypothesis will be a mapping assigning a Borel set of reals to each combination of a significance level with a parameter. To find a test statistic and a critical region in the case of the likely and suspicious implications, the following easily provable lemma will be useful.

Lemma 1. *Let* $\alpha \in (0, \frac{1}{2})$, $p \in (0, 1)$, $r_M \in \mathcal{N}$, *and* a *be a random variable obeying a binomial distribution with parameters* r_M *and* p. *Let further for each* $x \in \Re$, *the notation* $\lfloor x \rfloor$ *means the integer part of* x. *Denote*

$$K_a^{(!)}(\alpha, p) \stackrel{\text{def}}{=} \left\{ s : s \in \{0, \dots, r_M\} \ \& \ \sum_{i=s}^{r_M} \binom{r_M}{i} p^i (1-p)^{r_M - i} \leq \alpha \right\},$$
$$(11.23)$$

$$K_a^{(?)}(\alpha, p) \overset{\text{def}}{=} \left\{ s : s \in \{0, \ldots, r_M\} \ \& \ \sum_{i=0}^{s} \binom{r_M}{i} p^i (1-p)^{r_M - i} \leq \alpha \right\}.$$

(11.24)

Then it holds:

(a) $\displaystyle \max_{s \in \{0, \ldots, r_M\}} \Pr(a = s) = \Pr(a = \lfloor p(r_M + 1) \rfloor),$

(b) $(\forall s \in \{0, \ldots, \lfloor \frac{r_M}{2} \rfloor\}) \ \dfrac{\Pr(a = s)}{\Pr(a = r_M - s)} = \left(\dfrac{1-p}{p} \right)^{r_M - 2s},$

(c) *the fact that the random variable a assumes values in the set $K_a^{(!)}(\alpha, p)$ or in the set $K_a^{(?)}(\alpha, p)$ can be reformulated by means of the following equivalences:*

$$a \in K_a^{(!)}(\alpha, p) \quad \Leftrightarrow \quad \sum_{i=a}^{r_M} \binom{r_M}{i} p^i (1-p)^{r_M - i} \in (0, \alpha),$$

(11.25)

$$a \in K_a^{(?)}(\alpha, p) \quad \Leftrightarrow \quad \sum_{i=0}^{a} \binom{r_M}{i} p^i (1-p)^{r_M - i} \in (0, \alpha).$$

(11.26)

Provided we consider the null hypothesis '$p_{\psi|\varphi}$ is low', the parts (a)–(b) of this lemma in connection with (11.13) imply that under the validity of the null hypothesis we can expect a to assume most probably small values and the probability of a assuming a value $i < \dfrac{r_M}{2}$ to prevail over the probability of a assuming the complementary value $r_M - i$, this prevalence being the greater the smaller is i. That suggests using a one-sided binomial test which rejects the null hypothesis for high values of a. Similarly, if the null hypothesis '$p_{\psi|\varphi}$ is high' is considered, analogous reasons support the use of a one-sided binomial test rejecting the null hypothesis for low values of a.

If directly the random variable a would be used as a test statistic, then according to (11.12) the critical region for the level of significance α would be the set $K_a^{(!)}(\alpha, p)$ from (11.23) in the case '$p_{\psi|\varphi}$ is low', and the set $K_a^{(?)}(\alpha, p)$ from (11.24) in the case '$p_{\psi|\varphi}$ is high'. However, due to the part (c) of the lemma 11.3, the same results can be obtained using for the null hypothesis '$p_{\psi|\varphi}$ is low' the test statistic defined

$$T_{(!)}(p) \overset{\text{def}}{=} \sum_{i=a}^{r_M} \binom{r_M}{i} p^i (1-p)^{r_M - i} \quad | \quad p \in (0, 1),$$

(11.27)

for the null hypothesis '$p_{\psi|\varphi}$ is high' the test statistic defined

$$T_{(?)}(p) \overset{\text{def}}{=} \sum_{i=0}^{a} \binom{r_M}{i} p^i (1-p)^{r_M - i} \quad | \quad p \in (0, 1),$$

(11.28)

and in both cases the critical region defined

$$K(\alpha, p) \overset{\text{def}}{=} (0, \alpha) \quad | \quad \alpha \in (0, \tfrac{1}{2}), p \in (0, 1).$$

(11.29)

We shall use the latter setting (11.27)–(11.29), since then the critical region is independent of p, and also since it corresponds to the way how binomial tests are performed in the existing implementations of GUHA.

11.4   Fuzzy-Implicational Quantifiers

Definition 11.3 *Let $\rightarrow$ be a binary generalized quantifier of a monadic OPC, Π be a set of parameters, and $\tilde{\Pi}$ be a particular normalized fuzzy set on Π, i.e.*

$$\tilde{\Pi} = \{(p, \mu(p)) : p \in \Pi\} \ \text{where} \ \mu : \Pi \rightarrow \langle 0, 1 \rangle. \tag{11.30}$$

Then $\rightarrow$ is called fuzzy-implicational *w.r.t. $\tilde{\Pi}$ iff in addition to the associated function $\mathrm{Af}_\rightarrow$, a parametrized associated function $\mathrm{Paf}_\rightarrow : \mathcal{M}(2) \otimes \Pi \rightarrow \{0, 1\}$ is attached to it, with the following properties:*

- *$(\exists \pi \in \Pi) \ \mathrm{Paf}_\rightarrow(\cdot, \pi) = \mathrm{Af}_\rightarrow$,*

- *the functions $\mathrm{Faf}_\rightarrow^a : \mathcal{M}(2) \rightarrow \langle 0, 1 \rangle$ and $\mathrm{Faf}_\rightarrow^r : \mathcal{M}(2) \rightarrow \langle 0, 1 \rangle$, defined*

$$\mathrm{Faf}_\rightarrow^a(M) \overset{\text{def}}{=} \begin{cases} \sup\{\mu(\pi) : \pi \in \Pi \ \& \ \mathrm{Paf}_\rightarrow(M, \pi) = 1\}, \\ \qquad \text{if} \ M \in \mathcal{M}(2) \ \&(\exists \pi \in \Pi) \ \mathrm{Paf}_\rightarrow(M, \pi) = 1 \\ 0, \\ \qquad \text{if} \ M \in \mathcal{M}(2) \ \&(\forall \pi \in \Pi) \ \mathrm{Paf}_\rightarrow(M, \pi) = 0, \end{cases} \tag{11.31}$$

$$\mathrm{Faf}_\rightarrow^r(M) \overset{\text{def}}{=} \begin{cases} 1 - \sup\{\mu(\pi) : \pi \in \Pi \ \& \ \mathrm{Paf}_\rightarrow(M, \pi) = 0\}, \\ \qquad \text{if} \ M \in \mathcal{M}(2) \ \&(\exists \pi \in \Pi) \ \mathrm{Paf}_\rightarrow(M, \pi) = 0, \\ 1, \\ \qquad \text{if} \ M \in \mathcal{M}(2) \ \&(\forall \pi \in \Pi) \ \mathrm{Paf}_\rightarrow(M, \pi) = 1, \end{cases} \tag{11.32}$$

fulfil the condition

$$(\forall M_1, M_2 \in \mathcal{M}(2)) \ a_{M_2} \geq a_{M_1} \ \& \ b_{M_2} \leq b_{M_1} \Rightarrow$$
$$\Rightarrow \mathrm{Faf}_\rightarrow^a(M_2) \geq \mathrm{Faf}_\rightarrow^a(M_1) \ \& \ \mathrm{Faf}_\rightarrow^r(M_2) \geq \mathrm{Faf}_\rightarrow^r(M_1). \tag{11.33}$$

The function $\mathrm{Faf}_\rightarrow^a$ will be called an accepting fuzzy associated function *of $\rightarrow$ w.r.t. $\tilde{\Pi}$, $\mathrm{Faf}_\rightarrow^r$ will be called a* rejecting fuzzy associated function *of $\rightarrow$ w.r.t. $\tilde{\Pi}$.*

Remark 11.2 The fuzzy-implicationality of quantifiers is a generalization of their implicationality, in the sense that for $\tilde{\Pi}$ corresponding to a one-element crisp set, Definition 11.3 coincides with Definition 11.2.

Indeed, let $\pi_0 \in \Pi$, and let $\mathrm{Paf}_\rightarrow : \mathcal{M}(2) \otimes \Pi \rightarrow \{0, 1\}$ be an arbitrary function such that condition 11.3(i) holds (*e.g.* the cylindrical extension of $\mathrm{Af}_\rightarrow$ onto $\mathcal{M}(2) \otimes \Pi \otimes (0, 1)$). Then defining the membership function in

(11.30) by $\mu(\pi_0) \stackrel{\text{def}}{=} 1$ & $(\forall \pi \in \Pi \setminus \{\pi_0\}) \, \mu(\pi) \stackrel{\text{def}}{=} 0$, and taking into account (11.31)–(11.32) implies

$$(\forall M \in \mathcal{M}(2)) \; \text{Faf}^a_\rightarrow(M) = \text{Faf}^r_\rightarrow(M) = \text{Paf}_\rightarrow(M, \pi_0) = \text{Af}_\rightarrow(M), \tag{11.34}$$

which turns (11.33) into

$$(\forall M_1, M_2 \in \mathcal{M}(2)) \; a_{M_2} \geq a_{M_1} \; \& \; b_{M_2} \leq b_{M_1} \Rightarrow \text{Af}_\rightarrow(M_2) \geq \text{Af}_\rightarrow(M_1). \tag{11.35}$$

Due to the fact that $\text{Af}_\rightarrow$ can assume only the values 0 or 1, (11.35) is equivalent to (11.6).

Observe from (11.34) that both the accepting fuzzy associated function and the rejecting fuzzy associated function of a binary generalized quantifier coincide with its usual ($\{0, 1\}$-valued) associated function in the crisp case. That explains why one single concept of an associated function was sufficient to connect the notion of a generalized quantifier with common statistical hypotheses, though two concepts are needed to connect that notion with fuzzy hypotheses.

Theorem 11.1 *Let $\tilde{\Pi}$ be a normalized fuzzy set on $(0, 1)$, denote μ its membership function. Let further $M \in \mathcal{M}(2)$, a_M and r_M be the numbers assigned to M according to (11.5), and $T_{(!)}$, $T_{(?)}$ be realizations of the test statistics introduced in (11.27)–(11.28), obtained through replacing the random variable a with its realization a_M. Finally, let K be the critical region defined in (11.29), and $\alpha \in (0, \frac{1}{2})$.*
Then it holds:

(a) *Provided the conditions (11.19) are valid for μ, the generalized quantifier $\rightarrow^!$ is fuzzy-implicational w.r.t. $\tilde{\Pi}$, and its fuzzy associated functions w.r.t. $\tilde{\Pi}$ fulfil*

$$\text{Faf}^a_{\rightarrow^!}(M) = 1, \tag{11.36}$$

$$
\begin{aligned}
a_M > 0 \Rightarrow (\exists p^M_{(!)} \in (0, 1)) \; T^{-1}_{(!)}(\alpha) &= \{p^M_{(!)}\} \; \& \; \text{Faf}^r_{\rightarrow^!}(M) \\
&= 1 - \lim_{p \to p^M_{(!)}+} \mu(p). \quad (11.37)
\end{aligned}
$$

If in addition μ is right-continuous on $(0, 1)$, (11.37) simplifies to

$$
\begin{aligned}
a_M > 0 \Rightarrow (\exists p^M_{(!)} \in (0, 1)) \; T^{-1}_{(!)}(\alpha) &= \{p^M_{(!)}\} \; \& \; \text{Faf}^r_{\rightarrow^!}(M) \\
&= 1 - \mu(p^M_{(!)}). \quad (11.38)
\end{aligned}
$$

(b) *Provided the conditions (11.22) are valid for μ, the generalized quantifier $\rightarrow^?$ is fuzzy-implicational w.r.t. $\tilde{\Pi}$, and its fuzzy associated functions* *w.r.t.* $\tilde{\Pi}$ *fulfil*

$$a_M > 0 \Rightarrow (\exists p_{(?)}^M \in (0,1)) \, T_{(?)}^{-1}(\alpha) \;=\; \{p_{(?)}^M\} \; \& \; \mathrm{Faf}_{\to ?}^a(M)$$

$$=\; \lim_{p \to p_{(?)}^M -} \mu(p), \qquad (11.39)$$

$$\mathrm{Faf}_{\to ?}^r(M) = 0. \qquad (11.40)$$

If in addition μ *is left-continuous on* $(0,1)$, *(11.39) turns to*

$$a_M > 0 \Rightarrow (\exists p_{(?)}^M \in (0,1)) \, T_{(?)}^{-1}(\alpha) \;=\; \{p_{(?)}^M\} \; \& \; \mathrm{Faf}_{\to ?}^a(M)$$

$$=\; \mu(p_{(?)}^M). \qquad (11.41)$$

The proof of this theorem, as well as the proof of the following Theorem 11.2 can be found in [7].

Observe that the usual null hypotheses (11.14), corresponding to the likely implication, and (11.15), corresponding to the suspicious implications, can be viewed as fuzzy null hypotheses with the membership functions in Figures 11.1c) and 11.3c), respectively. These functions are not membership functions of singletons, hence the coincidence (11.34) is not valid. Nevertheless, the following theorem shows that the non-constant one from the two fuzzy associated functions of the quantifier coincides with its usual associated function in that case. Therefore, it is immaterial whether a likely/suspicious implication is given by means of a threshold, or by means of an interval delimited by that threshold and viewed as a specific fuzzy set. Moreover, the theorem establishes a connection between the values $p_{(!)}^M$, $p_{(?)}^M$ for $M \in \mathcal{M}(2)$, and the values which the associated functions of those two quantifiers assume for M.

Theorem 11.2 *Let* $\to \in \{\to^!, \to^?\}$, *and* $\theta \in (0,1)$. *Denote*

$$\tilde{\Pi}_{(!)} \;\overset{\mathrm{def}}{=}\; \{(p, \mu(p)) : p \in (0,1) \; \& \; \mu|\langle 0,\theta\rangle = 1 \; \& \; \mu|(\theta,1) = 0\}, \quad (11.42)$$

$$\tilde{\Pi}_{(?)} \;\overset{\mathrm{def}}{=}\; \{(p, \mu(p)) : p \in (0,1) \; \& \; \mu|\langle 0,\theta) = 0 \; \& \; \mu|\langle\theta,1) = 1\}. \quad (11.43)$$

Finally, let $p_{(!)}^M$ *and* $p_{(?)}^M$ *for* $M \in \mathcal{M}(2)$ *be the parameter values introduced in Theorem 11.1.*

Then:

(a) *provided* $\to \; = \; \to_\theta^!$, *the rejecting fuzzy associated function* $\mathrm{Faf}_\to^r$ *of* $\to$ *w.r.t.* $\tilde{\Pi}_{(!)}$ *coincides with* $\mathrm{Af}_\to$, *and they fulfil*

$$(\forall M \in \mathcal{M}(2)) \; \mathrm{Faf}_\to^r(M) = \mathrm{Af}_\to(M) = \begin{cases} 1 & \textit{iff} \quad p_{(!)}^M \geq \theta \\ 0 & \textit{iff} \quad p_{(!)}^M < \theta; \end{cases} \quad (11.44)$$

(b) *provided* $\to = \to^?_\theta$, *the accepting fuzzy associated function* $\text{Faf}^a_\to$ *of* $\to$
w.r.t. $\tilde{\Pi}_{(?)}$ *coincides with* $\text{Af}_\to$, *and they fulfil*

$$(\forall M \in \mathcal{M}(2)) \ \text{Faf}^a_\to(M) = \text{Af}_\to(M) = \begin{cases} 1 & \textit{iff} \ \ p^M_{(?)} > \theta \\ 0 & \textit{iff} \ \ p^M_{(?)} \le \theta. \end{cases} \quad (11.45)$$

11.5 Conclusion

In this chapter, a fuzzy-set approach to exploratory data analysis has been outlined, based on the general unary hypotheses automaton and on fuzzy hypotheses testing. This approach tries to bridge the gap between precise conditions on the distributions of the underlying random variables, typically required by common statistical tests, and the vagueness of the user's *a priori* knowledge, often accompanying exploratory data analysis. To this end, a fuzzy generalization of the concept of an implicational quantifier was proposed, and elaborated for the two GUHA implications usually encountered in practice, *i.e.* the likely implication and the suspicious implication.

The results presented in this chapter show that the usual method of generating GUHA implications is actually a special case of the proposed fuzzy approach, *i.e.* that approach can be indeed considered a fuzzy generalization of the traditional GUHA. However, it is only the basic principle of this generalization which is outlined here. To become practically applicable, the approach must be extended to cover a number of other concepts and issues.

Two such extensions, to GUHA associations and to the concept of a power function, are actually already the matter of an ongoing research. However, extending the fuzzy-set approach to GUHA associations implies facing a serious additional problem. Two of the commonly used types of associational quantifiers, namely the quantifier of a simple association, and the χ^2 quantifier, are actually based on asymptotic tests [1, 3, 4], and it is not at all apparent whether fuzzy hypotheses testing makes any sense in connection with asymptotic tests. Indeed, in the case of exact tests this approach attempts to introduce some vagueness into the exact distribution of the test statistic. However, asymptotic tests already implicitly contain a great amount of such vagueness since we actually do not know what the distribution of the test statistic is, knowing only that it weakly converges to a known asymptotic distribution if the sample size growths to infinity.

Naturally, the vagueness introduced into the distribution with fuzzy hypotheses testing represents quite another sort of vagueness than that inherent to an asymptotic test. The former can be viewed as a controlled vagueness, due to the possibility to determine the grade in which the test statistic is governed by a particular distribution from a parametrized set of probability distributions, through choosing a membership function on the parameter space. On the other hand, with an asymptotic test we do not have any control over the distribution of the test statistic. In fact, the set of all possible distributions cannot even be parametrized by a finite-dimensional parameter space. Since this feature pertains to the actual distributions, not to the asymptotic ones, it cannot be avoided through fuzzifying the parameter space of the asymptotic distribution. Therefore, fuzzifying the parameter space in the case of an asymptotic test, we would combine the controlled vagueness introduced by the approach with the uncontrolled vagueness inherent to the test into a new kind of vagueness which would again be uncontrolled, thus it would not allow both types of vagueness to be seen separately.

Intended future extensions of the approach will include at least foundedness, restricted sentences and missing data. From the effectiveness point of view, it is important to extend the approach to the concepts of hopeless antecedents and succedents, and to the notion of improving predicates.

Acknowledgements

Although this chapter is concerned with an application of fuzzy set theory, it also heavily relies on the theoretical background of the highly sophisticated GUHA approach. Therefore, I feel deeply obliged both to Petr Hájek and the late Tomáš Havránek, whose book [3] still remains the most complete presentation of that theoretical background. Moreover, it was Petr Hájek who caused me to start the work on fuzzy generalizations of GUHA, and who supported this work through valuable comments.

References

[1] Hájek P. (1984) The new version of the GUHA procedure ASSOC (generating hypotheses on associations) – mathematical foundations. In *COMPSTAT 1984 — Proceedings in Computational Statistics* 360–365.

[2] Hájek P. and Havránek T. (1977) On generating of inductive hypotheses. *International Journal of Man Machine Studies* 9: 415–438.

[3] Hájek P. and Havránek T. (1978) *Mechanizing Hypothesis Formation.* Springer-Verlag, Berlin.

[4] Hájek P., Sochorová A. and Zvárová J. (1995) GUHA for personal computers. *Computational Statistics and Data Analysis* 19: 149–153.

[5] Havránek T. (1980) An alternative approach to missing information in the GUHA method. *Kybernetika* 16: 145–155.

[6] Havránek T. (1980) Some comments on GUHA procedures. In *Exploratory Data Analysis*. Springer-Verlag, Berlin.

[7] Holeňa M. (1995) Fuzzy hypotheses for GUHA implications. Technical report, University of Paderborn, Department of Informatics – Cadlab.

[8] Rauch J. (1978) Some remarks on computer realization of the GUHA method. *International Journal of Man Machine Studies* 10: 75–86.

[9] Sochorová A. and Havránek T. (1990) A new version of the GUHA method for analysing categorial data. In *Computers in Medicine and Health Care* 169–170.

[10] Watanabe N. and Imaizumi T. (1993) A fuzzy statistical test of fuzzy hypotheses. *Fuzzy Sets and Systems* 53: 167–178.

[11] Yager R.R. (1992) On a semantics for neural networks based on fuzzy quantifiers. *International Journal of Intelligent Systems* 7: 765–786.

12

Fuzzy Logic Controller Based on Standard Operational Amplifiers

D. Kovačević, A. Kovačević

12.1 Introduction

Complex system modelling and control is really a challenging task where a lot of theories have been tested so far with more or less success. One of the successful theories has certainly been the fuzzy set theory because it proved itself not only by simulations, but also with numbers of real applications. In the last couple of years special attention has been given to a development of the special hardware supporting fuzzy logic and fuzzy set theory. In these attempts fuzzy control and fuzzy controllers are of great importance. Two approaches are possible: to adapt conventional digital computers for fuzzy control tasks or to develop dedicated fuzzy control hardware (devices) using either digital or analogue approaches. In the analogue approach two methods are possible; one is dealing with design and integration of original circuits, while the other concerns standard devices, either transistors or ICs, in discrete form. On the other hand, the applications of fuzzy set theory have entered into every day practice. Therefore, it was necessary to introduce related facts in the domain mentioned through appropriate innovations in faculty curricula. Naturally, many different approaches are possible, but for the sake of technological and economic reasons, we have been obliged to accept an approach based on conventional, non-expensive devices such as general purpose operational amplifiers, multipliers, multiplexers, transistors, etc.

An operational amplifier (Op-Amp) is the most extensively used analogue integrated circuit. The objective of this chapter is to describe the possibility

FUZZY LOGIC
Editor J. F. Baldwin

of applying Op-Amp as a fuzzy hardware supporting tool.

The definitions of the operations on fuzzy sets we provide here are so-called *default definitions* which are used most often. All those operations which are extensions of crisp concepts reduce to their usual meaning when the fuzzy subsets have membership grades that are drawn from {0,1}. For this reason, when extending operations to fuzzy sets we use the same symbols as in a set theory.

We shall introduce circuits based on Op-Amps supporting the following operations: union (MAX), intersection (MIN) and complement (negation). Of course, if one is going to design (build) a fuzzy logic controller then the design of a membership function circuit (MFC) is unavoidable [1]. Sometimes, a membership function generator (MFG) is necessary, too [1]. Their tasks are to generate discrete (MFG) or continuous membership functions and to give, if necessary, the degree of its satisfaction for (normalized) input signals (MFC). Besides these circuits, membership functions circuits with the possibility of changing the maximum grade of membership by external applied voltage are presented in this chapter. Such circuits are called grade-controllable membership function circuits (GCMFC) [2], and they are used in the design of singleton fuzzy logic controllers (SFLC). All these circuits are based on existing (standard) [3] IC-components and/or exclusively on general purpose operational amplifiers.

12.2   Operational Amplifiers Support Operations on Fuzzy Sets

12.2.1   Union

Definition 12.1 *Assume A and B are two fuzzy subsets of X. Their union is a fuzzy subset C of X, denoted $C = A \cup B$, such that for each $x \in X$*

$$C(x) = \max[A(x), B(x)] = A(x) \cup B(x). \tag{12.1}$$

In the fuzzy set literature it is a common practice to use $\cup$ as the max *operator* [4]. *The operation* max *can be written in algebraic terms* [4]:

$$\max(a, b) = [(a + b) + \mid a - b \mid]/2. \tag{12.2}$$

Example 12.1 Let $X = \{x_1, x_2, x_3, x_4, x_5\}$. Assume A and B are two fuzzy subsets of X where

$$A = \{1/x_1, 0.7/x_2, 0.3/x_3, 0/x_4, 0.9/x_5\}$$

and

$$B = \{0.2/x_1, 0.9/x_2, 0.4/x_3, 1/x_4, 0.4/x_5\}.$$

Then

$$C = \{1/x_1, 0.9/x_2, 0.4/x_3, 1/x_4, 0.9/x_5\}.$$

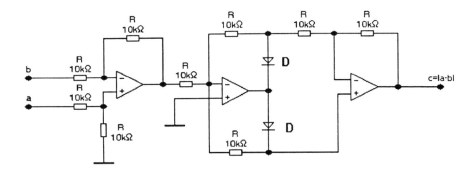

Figure 12.1: Circuit generating absolute value of difference (CGAVD) — version 1.0.

The first element of subset C can be evaluated from relation (12.2):

$$\max(1/x_1, 0.2/x_1) = (1 + 0.2 + |1 - 0.2|)/2 = 1.$$

As all other elements of subset C can be evaluated in the same manner, we can conclude about the necessity of transferring relation (12.2) into an appropriate hardware form. It is obvious that an adder in combination with an absolute-value circuit [5], as shown in Figure 12.1, can be used to fit (functional) relationship (12.2).

Another version of the absolute-value circuit (CGAVD) is shown in Figure 12.2. With this circuit one can generate an absolute value of the difference between two input voltages V_1 (a) and V_2 (b). The circuit handles input signals over the voltage range of 0 to +5V and operates from a single +5V power supply. Its output voltage is equal to $5 |V_1 - V_2|$ (V). Adding a (non) inverting amplifier with transfer function $|G| = 1/5$, necessary matching is achieved, and in such an arrangement this circuit can be used in circuits performing either MIN or MAX operation.

In the design of a MAX operation circuit the output of the CGAVD circuit (see Figure 12.1) feeds one of the three inputs of the adder. Voltages V_a and V_b corresponding to algebraic terms a and b, respectively, are applied to two other inputs of the adder (see Figure 12.3). The output (V_o) is given by the equation:

$$V_o = \frac{R_1 + R_2}{R_1}(1/n)(V_a + V_b + V_{|a-b|}), \tag{12.3}$$

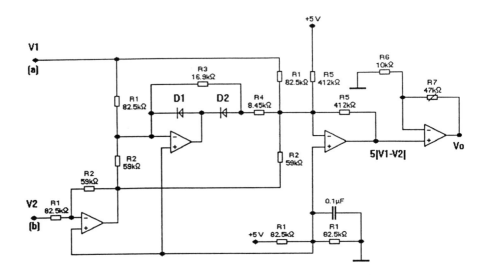

Figure 12.2: Practical realization of CGVD — version 2.0.

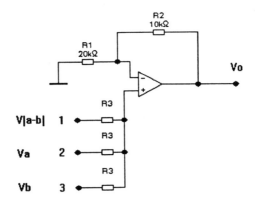

Figure 12.3: Adder (as a part of MAX-circuit).

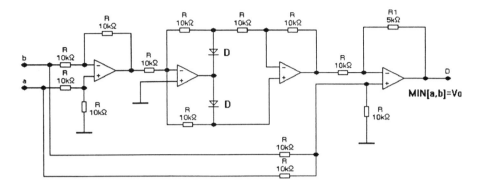

Figure 12.4: MIN-circuit.

Table 12.1: A part of experimental results (MIN-circuit).

U_1	0,97	0,98	1,00	1,03	1,08	V
U_2	1,08	2,03	3,01	3,99	4,98	V
U_{out}	0,99	0,97	0,96	1,01	1,05	V

U_1	1,98	2,00	2,02	2,03	2,05	V
U_2	1,03	2,04	3,04	4,04	5,08	V
U_{out}	1,03	1,97	1,98	1,97	1,99	V

U_1	2,99	3,00	3,01	3,03	3,04	V
U_2	1,01	2,00	3,01	4,02	5,06	V
U_{out}	0,98	1,97	2,98	3,00	2,99	V

U_1	3,97	3,99	3,99	4,02	4,04	V
U_2	1,03	2,04	3,02	4,05	4,97	V
U_{out}	0,98	2,00	3,02	4,01	3,99	V

where n is a number of equal resistors (R_3) connected to the noninverting input of the adder. Choosing adequate values for resistors R_1 and R_2 in equation (12.3): $R_1 = 20K$ and $R_2 = 10K$ hardware realization of the MAX operation is completed.

12.2.2 Intersection

Definition 12.2 *Assume A and B are two subsets of X. Their intersection is a fuzzy subset D of X, denoted $D = A \cap B$, such that for each $x \in X$*

$$D(x) = \min[A(x), B(x)] = A(x) \wedge B((x). \tag{12.4}$$

It is common practice in the fuzzy literature to use $\wedge$ as the MIN operator.

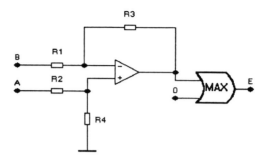

Figure 12.5: Relative complement circuit.

The operation MIN can be written in algebraic form:

$$\min(a, b) = (a + b - |a - b|)/2. \tag{12.5}$$

Concerning the same approach in designing a hardware circuit as in the previous paragraph, we developed a circuit performing MIN operation (see Figure 12.4). Experimental results are presented in Table 12.1.

12.2.3   Complement

Now we extend the operations of relative complement and complement (or negation) to fuzzy subsets.

Definition 12.3 *Assume A and B are two fuzzy subsets of X. The relative complement of B with respect to A, denoted $E = A - B$, is defined as the fuzzy subset E of X, where for each $x \in X$:*

$$E(x) = \max[0, A(x) - B(x)]. \tag{12.6}$$

Example 12.2 Assume

$$A = \{0.5/a, 0.3/b, 0.7/c, 0.6/d, 1/f\}$$

and

$$B = \{0.5/a, 1/b, 0.3/c, 0/d, 0.5/f\}$$

then

$$E = A - B = \{0/a, 0/b, 0.4/c, 0.6/d, 0.5/f\}.$$

A combination of a differential amplifier and the MAX-circuit, we have just designed, performs the operation of relative complement (see Figure 12.5).

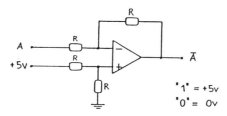

Figure 12.6: Complement circuit.

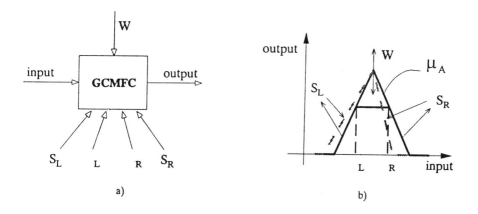

Figure 12.7: GCMFC (a) and its function (b).

Definition 12.4 *Assume A is a fuzzy subset of X. The complement or negation of A denoted $\bar{A}$, is defined as the fuzzy subset $\bar{A} = X - A$. Hence, for each $x \in A$*

$$\bar{A}(x) = 1 - A(x). \tag{12.7}$$

Thus the negation is the complement of A with respect to the whole space X. Hardware realization based on operational amplifiers is simple: the differential amplifier shown in Figure 12.6 gives desired voltage on its output.

12.3 Grade Controllable Membership Function Circuits

In fuzzy hardware, a membership function circuit is considered as an electronic circuit that can produce the compatibility of the input signal to the label in

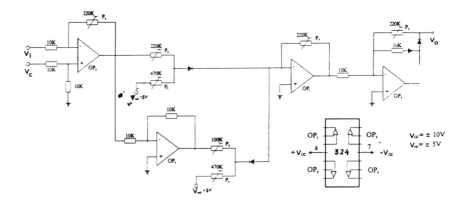

Figure 12.8: GCMFC constructed with general purpose amplifiers.

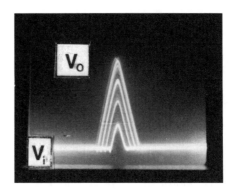

Figure 12.9: Input-output characteristics of GCMFC.

the control rule.

It is common practice to adopt a piecewise linear membership function (μ_A). The membership function is characterized with five parameters: U_L, U_R, S_L, S_R and W. S_L and S_R are the slopes of the left-hand side and right-side of the membership function, respectively. Reference voltages V_L and V_R define position and the shape of the membership function as well. A maximum value of the grade of membership (W) means the weighting of the MFC (see Figure 12.7). This parameter is essential for defuzzification without the algebraic division. A practical circuit realization is shown in Figure 12.8. Potentiometers P_1, P_4, P_5 and P_6 are used for the membership function shaping, and its

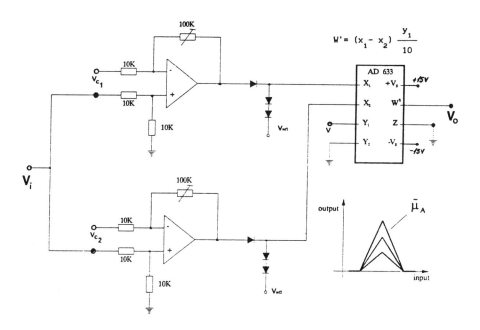

Figure 12.10: GCMFC based on a low cost analogue multiplier AD 633.

relative position is controlled by voltage input U_{cont}. The input-output characteristic of such a MFC can be changed (grade of membership) by applied voltage U_W in the way shown in Figure 12.9. Therefore, this circuit is called a grade-controllable membership function circuit [2, 6].

Another version of GCMFC is shown in Figure 12.10. Grade controllability is achieved by voltage applied at Y_1-input of the analogue multiplier AD 633. W' is the transfer function of the AD 633, as indicated in Figure 12.10.

Both circuits described above can be used in designing a singleton-consequent fuzzy logic controller, because a (voltage) feedback loop can be used for controlling the (maximum) grade of membership. At the same time, these GCMFCs can be used as the building blocks of an ordinary analogue fuzzy logic controller [7, 8].

12.4   SFLC

In this section a singleton-consequent fuzzy logic controller (SFLC) to support the teaching of fuzzy logic circuits at the Maritime Faculty Dubrovnik Department Split will be presented [9, 10].

In a control system, a fuzzy logic controller (FLC) accepts deterministic information and produces deterministic signal(s) to actuate the system under control. In an ordinary FLC, the variable in the consequent is represented with a fuzzy linguistic value. Defuzzification is very complicated because the FLC needs a weighted sum divided by summation of the inference result to get the centre of gravity. In the traditional fuzzy inference engine or fuzzy rule chip, a fuzzy word in the consequent is constructed with seven (five) non-zero elements. The smallest number of non-zero elements in the consequent is one, and in that case we are talking about SFLC [2]. We used standard operational amplifier technology, and the method which gives a centre of gravity of the inference result without an analogue divider. Our SFLC utilizes a set of grade-controllable membership function circuits (GCMFC), a set of MIN-circuits, a weighted adder and an ordinary adder in conjunction with a matching amplifier (see Figure 12.11). Note that the realization of the MIN (MAX) circuit can be based on a diode concept as it was introduced in the early days of classical digital technic. In that case, Op-Amp is used as a buffer and for compensation of an error (offset voltage – V).

Three control rules are implemented. The antecedent of each rule includes three variables e, Δe and $\Delta^2 e$. The consequent of each rule includes one singleton variable c or one deterministic value C.

$$C' = \frac{\sum_{i=1}^{3} C_i \{\mu_{e_i}(e') \wedge \mu_{\Delta e_i}(\Delta e') \wedge \mu_{\Delta^2 e_i}(\Delta^2 e')\}}{\sum_{i=1}^{3} \mu_{e_i}(e') \wedge \mu_{\Delta e_i}(\Delta e') \wedge \mu_{\Delta^2 e_i}(\Delta^2 e')} \tag{12.8}$$

Control inputs are applied to GCMFCs to produce compatibilities of the inputs to the labels. These compatibilities feed three-input MIN-circuits. MIN's outputs are summed twice; once they are summed in the inverting

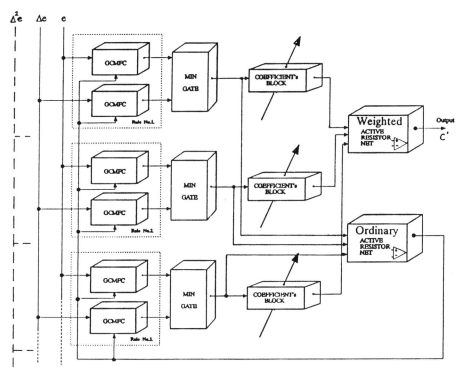

Rule No.1: If e=PM and Δe=PS then z=C_1
Rule No.2: If e=PM and Δe=NO then z=C_2
Rule No.3: If e=PS and Δe=NO then z=C_3

Figure 12.11: Architecture of a SFLC.

input of an operational amplifier, while the unit voltage (5V) is applied to
the non-inverting input of the same amplifier at the same time (inside of the
ordinary active resistor net). The output of that amplifier, in conjunction
with a matching amplifier, controls the common weights of all GCMFCs so
that summation of the outputs of MIN-circuits is always equal to unity (see

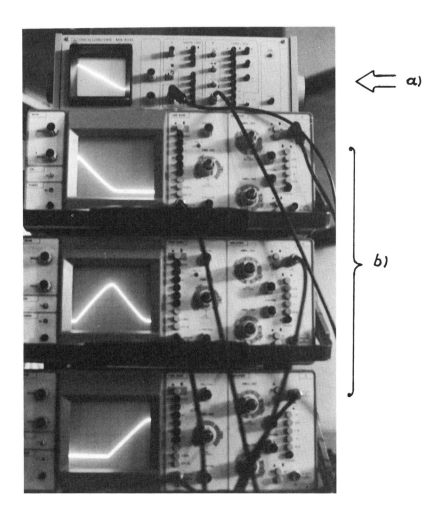

Figure 12.12: Input-output characteristics of the SFLC (a); a set of control rules (b).

Figure 12.11). On the other hand, outputs of MIN-circuits are weighted by means of a resistors net and are summed (operational amplifier) to give the output of the SFLC. This output represents the centre of gravity of the inference result, since the denominator in equation (12.8) is always equal to unity. The SFLC was tested using three control rules, each of which includes a fuzzy value in the antecedent and a deterministic value in the consequent. These three control rules and the experimental result (output of the SFLC) are shown in Figure 12.12. The procedure of setting the SFLC is illustrated in Figure 12.13.

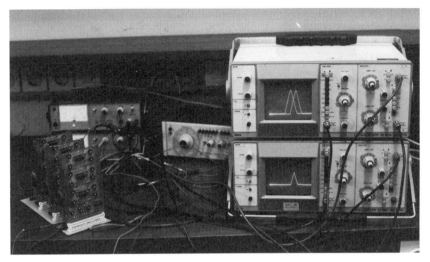

Figure 12.13: Setting the SFLC.

12.5   Conclusion

Operational amplifiers can be used as a platform for the design of circuits supporting operations on fuzzy sets. The membership function (MFC) as well as fuzzy set(s) (MFG) can be generated through circuits based on Op-Amp. In that way the field of the operational amplifiers' application is extended to the domain of the new theory, showing the versatility of Op-Amps at the same time. Resolving the problem of hardware support based on Op-Amps to fuzzy set theory operations, it is possible to design a fuzzy logic controller using exclusively, or at least mostly, general purpose operational amplifiers. In the other case, low cost multipliers and multiplexers are necessary as background for Op-Amp's platform. The basic advantages of this approach in designing (S)FLC can be summarized as follows:

- simple circuitry,

- no special hardware design (devices) is (are) necessary,

- simple definition terms are easy to obtain,

- comprehensible,

- student (education) oriented,

- inexpensive,

- striking clearness,

- can be regarded as the first but powerful step in understanding the notion and design of VLSI fuzzy chips,

- both controllers, with and without divider are realizable,

- many useful laboratory experiments at an undergraduate level can be performed.

References

[1] Yamakawa T. (1989) Stabilization of an Inverted Pendulum by a High-Speed Fuzzy Logic Controller Hardware System. *Fuzzy Sets and Systems* 32: 161–180.

[2] Yamakawa T. (1989) An Application of a Grade - Controllable Membership Function Circuit to a Singleton-Consequent Fuzzy Logic Controller. *Proc. of the Third IFSA Congress* 269–305.

[3] Kovačević A. *et al.* (1994) Hardware Support to Fuzzy Set Theory Operations. *Proceedings of MIPRO'94 - MEET* 2-92–2-97.

[4] Yager R.R. and Filev D.P. (1994) Operations on Fuzzy Sets. *Essentials of Fuzzy Modelling and Control*, 2–12, John Wiley & Sons.

[5] *Analog Devices Nonlinear Circuits Handbook*, (1976) 361–363.

[6] Ishizuka O. *et al.* (1992) Design of a Fuzzy Controller with Normalization Circuits. *Proc. of the 5th IFSA World Congress* 1303–1308.

[7] Yamakawa T. (1986) High Speed Fuzzy Controller Hardware System. *Proc. of the 2nd Fuzzy System Symposium* 122–130.

[8] Kovačević D. and Stipaničev D. (1993) Hardware Realization of Fuzzy Set Membership Function Generator. *Proceedings of MIPRO'93 - MEET* 4-36–4-41.

[9] Kovačević D. *et al.* (1993) New Trends in Electronics Teaching. *Proceedings of MIPRO'93 - MEET* 4-24–4-29.

[10] Kovačević D. *et al.* (1994) Laboratory Experiments with Fuzzy Set Modules. *Proceedings of KOREMA* 431–435.

13

Fuzzy Sets and Community Transport

R.I. John, S.C. Bennett

13.1 Introduction

Fuzzy set theory has been applied to vehicle routing and scheduling problems, most notably by Kikuchi and collaborators. Teodorovic and Kikuchi [7] apply fuzzy set theory to the selection of tours for the single depot multiple delivery routeing problem. Kikuchi and Donnelly [5], Kagaya, Kikuchi and Donnelly [4] and Kikuchi and Vukadinovic [6] explore the application of fuzzy set theory to scheduling demand-responsive transport, typically dial-a-ride operations. This chapter presents the results of applying fuzzy techniques to community transport (CT).

In the United Kingdom, CT operators provide a range of minibus based services for voluntary and community organizations, including 'group hire' services, in which the CT hires out vehicles to member groups. These vehicles may be specially equipped to carry passengers with disabilities, by means of tail-lifts or ramps and spaces for wheelchairs. In some areas, CT operators also provide a 'vehicle brokerage' service in which the CT acts as a broker between organizations which own vehicles, but do not use them all the time and groups which require transport but do not own vehicles. Warrington [8] and Warrington, Bryman and Gillingwater [9] have explored the background to and operation of these services.

Bennett [1] outlines the problems faced by vehicle brokerage operators in allocating vehicles to bookings. Two sets of criteria are applied by operators to the allocation of vehicles, availability criteria and sequencing criteria. Availability criteria are used to determine whether a vehicle is available, based on the diary system within the software, or suitable in terms of whether it has the seating or wheelchair capacity or meets other criteria for the trip. Sequencing criteria are then used to sort the vehicles into a specific order for presentation to the person taking the booking. Sequencing criteria are used to ensure that the most appropriate vehicles are considered first. One component attempts to measure how 'difficult to book' a vehicle is. The aim is to consider vehicles that are difficult to book first, so that they are matched with appropriate trips as early as possible. It is all too easy for booking staff to allocate trips to vehicles which they know are likely to be suitable and available, and then find that they cannot allocate trip bookings which are made later.

At the suggestion of Birmingham Shared Transport Services (BSTS), a 'sequence number' in the range 1 to 100 is assigned to each vehicle. This is a subjective measure of how difficult to book a vehicle is. It is difficult to assign values to the sequence number. 'Difficult to book' is a fuzzy concept that is dependent on other fuzzy factors such as accessibility and availability. Initial work [2] indicated that the application of fuzzy logic and fuzzy set theory to this problem is promising and that a bookability value can be determined for each vehicle that gives a measure of how difficult to book a vehicle is. This chapter is structured in the following way. Section 13.2 looks at the how the fuzzy rules were acquired and what was the outcome of this acquisition. Section 13.3 considers the approach that was adopted for combining the rules, the results are presented in section 13.4 and the conclusions in section 13.5.

13.2   Knowledge Acquisition and Representation

The co-ordinator and booking staff at CCT were interviewed to establish which factors they considered to affect the booking of vehicles in brokerage. A range of factors were established, and some of these were generalized to simplify the representation.

A number of variables were identified, and for each, initial fuzzy sets established. The actual sets adopted were arrived at by a combination of knowledge acquisition and investigation of historical data. They were:

- Accessibility (low, high);

- Well-equippedness (low, medium high);

- Own-use frequency (low, medium, high);

- Own-use extent (low, medium, high);

- Capacity (small, medium, large).

These were used to derive values for three intermediate fuzzy variables:

- Flexibility (low, medium, high);

IN USE

Own Use Extent	Own Use Frequency		
	L	M	H
L	L	L	M
M	L	M	H
H	M	H	H

FLEXIBILITY

Well Equipped	Accessible	
	L	H
L	L	M
M	L	H
H	M	H

USEABILITY

Capacity	Flexibility		
	L	M	H
L	L	M	M
M	L	M	H
H	M	M	H

BOOKABILITY

In-Use	Useability		
	L	M	H
L	L	H	VH
M	L	L	H
H	VL	L	L

Figure 13.1: The fuzzy rules.

- In-use (low, medium, high);

- Useability (low, medium, high).

These last two were then combined to give a single fuzzy set for Bookability:

- Bookability (very low, low, high, very high),

which is treated as the opposite of 'difficult to book'. This structure reflects the discussions with the expert. The fuzzy sets were then combined in fuzzy rules to give a knowledge base that can be represented by matrices. The initial complete set of rules (Figure 13.1) can be represented in the influence diagram (Figure 13.2).

Data was extracted from the CCT computerized booking system for the six months 1st July 1993 to 31st December 1993. The data was analysed

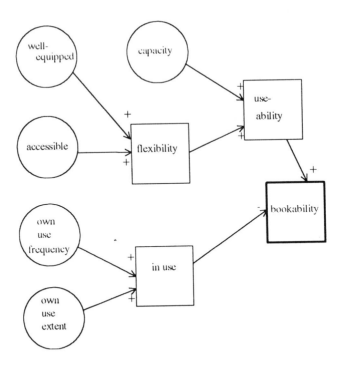

Figure 13.2: The influence diagram.

mainly to provide measures which could be used to determine the membership function for each vehicle in each of the fuzzy sets. Own-use frequency was calculated as the average frequency of trips per week over 26 weeks, while own-use extent was the average duration of own-use trips in hours per week. Other data was taken directly from the vehicle records in the computer system: accessibility in terms of the number of places for wheelchairs in the vehicle, and well-equippedness in terms of the number of different facilities available on the vehicle.

13.3   Fuzzy Composition

There are various methods available for working with fuzzy sets [3]. The process breaks down into the following phases:

- Define the fuzzy sets to be used;

- Aggregate the fuzzy sets;

- Decompose the aggregated sets;

- Defuzzify the solution.

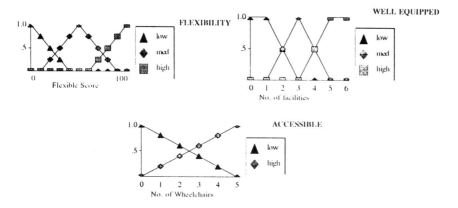

Figure 13.3: Three fuzzy sets.

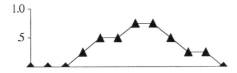

Figure 13.4: A typical truncated set.

The technique adopted in this work for composing the final set (the compositional rules of inference) is known as the min=max method. For each rule the consequent fuzzy region is the minimum of the truths. The output fuzzy region is the maximum of these minimized fuzzy sets.

So, for example, we have the fuzzy variables well-equipped, accessible, and flexibility defined by the diagrams in Figure 13.3.

An example rule in the system is 'if accessibility is high and well-equipped is low then flexibility is medium'. If, for example a particular vehicle had two facilities then the fuzzy value for the set *well-equipped is low* would, from the graph, be 0.5. If the same vehicle had three wheelchair spaces then *accessibility is high* would have a value of 0.6. The procedure is to take the minimum of these (0.5) and using the *flexibility is medium* set truncate the set.

This approach is repeated for each rule yet each time only taking the maximum of the truncated set, typically giving a shape as in Figure 13.4.

This set has now to be defuzzified by, in some way, arriving at a single figure that can be used for comparison purposes. There are various methods available for defuzzification. One of the most widely used techniques, which has been adopted here, is that of centroid defuzzification. This is essentially the weighted mean of the fuzzy region and is defined by:

$$\frac{\sum d\mu(d)}{\sum \mu(d)}$$

where d is the domain value for membership value $\mu(d)$.

13.4   Findings

Following the initial work it was felt that an important analysis that needed
to be carried out was to look at how the rules deployed affect the bookability.
There is no clear way of measuring how good a solution is other than to test
it on historical data and calculate possible savings in hours for vehicles and
number of vehicles used. Data supplied by Camden Community Transport
covered 36 vehicles over a six month period. The fuzzy rules and method
described above were used to arrive at a bookability value for each vehicle.
Initial work [2] had shown that this approach produced favourable results
compared with the way the CT operators carried out bookings. However, to
investigate how the rules affect the discrimination between cases the spread
of values and the ordering were compared. Hence the standard deviation,
rank correlation and mean of the bookability values were used to compare
approaches. Experiments were carried out using subsets of the rules. Three
approaches were considered:

The intermediate approach
     This approach uses the rules shown in Figure 13.1 and is called the inter-
mediate approach, since the basic fuzzy sets are used to arrive at intermediate
sets which, in turn, provide the Bookability set.

The simple approach
     Here we have only 14 simple rules that directly link the base sets to Book-
ability. An example rule might be

     IF Capacity is Low THEN Bookability is Low.

The all factors approach
     This approach adopts rules where Bookability is described explicitly using
all bases sets — a total of 162 rules. An example rule might be

          IF Accessibility is High;

          AND Well Equipped is High;

          AND Capacity is High;

          AND Own Use Extent is Low;

          AND Own Use Frequency is Low;

          THEN Bookability is Very High.

Table 13.1: The methods compared.

	mean	s.d.
Intermediate	47.8	8.36
Simple	50	4.36
All Factors	53	2.56

Table 13.2: Results of simulated booking with Camden data.

Vehicles Used	All Factors	Intermediate	Simple
145	19	21	20
290	30	30	31
TBAL Trips			
145	1	1	2
290	18	14	17
TBAL Hours			
145	8	8	164
290	630	613	622

The standard deviation and mean for each method are shown in Table 13.1. What was found was surprising:

(a) Using more rules did not necessarily give a better spread of values, in fact the initial, intermediate approach gave the best spread.

(b) Despite the fact that it was thought that the rules carried the same implications about the domain, the rank correlations of the resulting ordering of vehicles indicated that they were producing radically different effects for some vehicles.

This led to consideration of the impact of the rules in more detail and, in particular, to measure how well the different rank orderings of vehicles produced by the three sets of rules work when applied to the bookings as mentioned above. To test this, a simulation of the booking process was run using the three different sets of rules and data based on the historical data from Camden CT. The results of this are shown in Table 13.2.

A sample of bookings for one week was used. This contained 145 bookings. The booking process was run for this data, and the number of bookings was then doubled to study how the system performed as the number of bookings was increased to saturate the vehicles.

The vehicles were presented to the booking algorithm in the rank order generated by each of the three methods of constructing the fuzzy rules (All

Factors, Intermediate and Simple). The booking algorithm takes each vehicle trip and attempts to allocate it to to each vehicle in turn, checking:

a) whether the vehicle is already in use at the times requested;

b) whether the vehicle has the facilities required by the users;

c) whether the vehicle has the seating capacity required by the users, and

d) whether the vehicle has the wheelchair capacity required by the users.

The table shows three figures for each run through the data with the vehicles in each of the three rank orders. These figures are:

- the number of vehicles used to allocate as many bookings as could be allocated;

- the number of trips which could not be allocated, and

- the duration in hours of the trips which could not be allocated

(TBAL stands for To Be Allocated Later).

As can be seen, there is little difference in the number of vehicles used, but the Intermediate rule set performs better than the other two in minimizing the number of trips and the duration of trips in hours which cannot be allocated automatically. In the case of the number of unallocated hours the simple approach is statistically significantly worse than the other two methods.

We conclude that the Intermediate rule set provides a better way of deriving a measure of how difficult to book the vehicles are.

13.5   Conclusions

CT operators currently assign a sequencing number to vehicles when scheduling bookings. This has proved to be unsatisfactory. This chapter reports the results of applying various sets of fuzzy rules to this problem. Bookability vehicles have been calculated for 36 vehicles used by Camden Community Transport. Various measures were used to assess the efficacy of the approaches.

In terms of discriminatory ability the rules where all possible combinations were employed was found to be very poor; the intermediate approach was good at discriminating but the ordering was radically different from the other techniques.

When the approaches were applied to actual data the simplest approach was found to be significantly the worst in terms of allocating trips and hours. The Intermediate approach which was modelled on the schedulers perceptions was found to marginally outperform the all factors approach. This is useful since this requires less rules.

On balance, the intermediate approach was best although further investigation will have to be carried out into the reasons for the different ordering for these set of rules.

Acknowledgements

Our thanks to Camden Community Transport for their co-operation and for access to their data.

References

[1] Bennett S.C. (1992) The Design and Development of an Information Technology Application for Community Transport: A Case Study Approach, Unpublished MPhil Thesis, Department of Transport Technology, School of Engineering, Loughborough University of Technology.

[2] Bennett S.C. and John R.I. (1994) The application of fuzzy set theory to vehicle brokerage. *International Symposium on Automotive Technology and Applications, Dedicated Conference on Mechatronics.* Aachen.

[3] Cox E. (1994) *The Fuzzy Systems Handbook.* AP Professional.

[4] Kagaya S., Kikuchi S. and Donnelly R.A. (1991) Use of Fuzzy Theory technique for Grouping of Trips in Vehicle Routing and Scheduling Problem. Paper accepted for publication in *European Journal of Operational Research.*

[5] Kikuchi S. and Donnelly R.A. (1992) Scheduling Demand-Responsive Transportation Vehicles using Fuzzy-Set Theory. *Journal of Transportation Engineering* (118) 3: 391–409.

[6] Kikuchi S. and Vukadinovic D. (1993) Grouping Trips by Fuzzy Similarity for Scheduling of Demand-Responsive Transportation Vehicles. Paper submitted to *Transportation Planning and Technology.*

[7] Teodorovic D. and Kikuchi S. (1991) Application of fuzzy sets theory to the saving based vehicle routing algorithm. *Civil Engineering Systems* (8) 87–93.

[8] Warrington A. (1986) The Co-ordination of Community Transport Resources: A Case Study of "Shared Transport Service", Birmingham, TT8603, Department of Transport Technology, School of Engineering, Loughborough University of Technology.

[9] Warrington A., Bryman A. and Gillingwater D. (1987) The Co-ordination of Community Transport Resources: A Case Study of the Minibus Pooling Scheme Administered by Tower Hamlets Community Transport, TT8701, Department of Transport Technology, School of Engineering, Loughborough University of Technology.

Index